Laboratory Methods in Microfluidics

Laboratory Methods in Microfluidics

Basant Giri
Center for Analytical Sciences,
Kathmandu Institute of Applied Sciences,
Kathmandu, Nepal

ELSEVIER

Elsevier
Radarweg 29, PO Box 211, 1000 AE Amsterdam, Netherlands
The Boulevard, Langford Lane, Kidlington, Oxford OX5 1GB, United Kingdom
50 Hampshire Street, 5th Floor, Cambridge, MA 02139, United States

Notices
Knowledge and best practice in this field are constantly changing. As new research and experience broaden our
understanding, changes in research methods, professional practices, or medical treatment may become necessary.

Practitioners and researchers must always rely on their own experience and knowledge in evaluating and using any
information, methods, compounds, or experiments described herein. In using such information or methods they
should be mindful of their own safety and the safety of others, including parties for whom they have a professional
responsibility.

To the fullest extent of the law, neither the Publisher nor the authors, contributors, or editors, assume any liability
for any injury and/or damage to persons or property as a matter of products liability, negligence or otherwise, or
from any use or operation of any methods, products, instructions, or ideas contained in the material herein.

British Library Cataloguing-in-Publication Data
A catalogue record for this book is available from the British Library

Library of Congress Cataloging-in-Publication Data
A catalog record for this book is available from the Library of Congress

ISBN: 978-0-12-813235-7

For Information on all Elsevier publications
visit our website at https://www.elsevier.com/books-and-journals

Working together
to grow libraries in
developing countries

www.elsevier.com • www.bookaid.org

Publisher: John Fedor
Acquisition Editor: John Fedor
Editorial Project Manager: Emily Thomson
Production Project Manager: Anitha Sivaraj
Designer: Miles Hitchen

Typeset by MPS Limited, Chennai, India

Contents

About the Author   xiii

Preface   xv

Acknowledgments   xvii

1	Introduction to Microfluidics	1
	1.1 Background	1
	1.2 Frequently Used Microfluidic Terms	2
	1.3 Assessment Questions	6
	References	7
2	Fabrication of a Glass Microfluidic Device	9
	2.1 Background	9
	2.2 Microfluidic Device Design	15
	2.3 Chemicals and Supplies	15
	2.4 Hazards	16
	2.5 Experimental Procedure	16
	2.6 Additional Notes	19
	2.7 Assessment Questions	20
	References	20
3	Fabrication of a Paper Microfluidic Device for Blood-Plasma Separation	21
	3.1 Background	21
	3.2 Microfluidic Device Design	23
	3.3 Chemicals and Supplies	24
	3.4 Hazards	24

3.5 Experimental Procedure 24

3.6 Additional Notes 26

3.7 Assessment Questions 26

References 26

4 Fabrication and Testing of a PDMS Microchip 27

4.1 Background 27

4.2 Design of the Microfluidic Device 27

4.3 Chemicals and Supplies 28

4.4 Hazards 28

4.5 Experimental Procedure 28

4.6 Additional Notes 31

4.7 Assessment Questions 31

References 32

5 Determination of Electroosmotic Flow in a Glass Microfluidic Device Using a Neutral Marker 33

5.1 Background 33

5.2 Microfluidic Device Design 34

5.3 Chemicals and Supplies 35

5.4 Hazards 35

5.5 Experimental Procedure 35

5.6 Additional Notes 37

References 37

6 Electrophoretic Separation in a Microchannel 39

6.1 Background 39

6.2 Design of the Microfluidic Device 40

6.3 Chemicals and Supplies 42

6.4 Hazards 42

6.5 Experimental Procedure 42

6.6 Additional Notes   45

6.7 Assessment Questions   45

References   45

7   A Simple Experiment for the Study of Droplet
     Microfluidics   47

7.1 Background   47

7.2 Microfluidic Device Design   48

7.3 Chemicals and Supplies   48

7.4 Hazards   48

7.5 Experimental Procedure   48

7.6 Additional Notes   50

7.7 Assessment Questions   50

References   50

8   Laminar Flow and Diffusion in a Microchannel   51

8.1 Background   51

8.2 Microfluidic Device Design   52

8.3 Chemicals and Supplies   52

8.4 Hazards   52

8.5 Experimental Procedure   53

8.6 Additional Notes   54

8.7 Assessment Questions   54

References   55

9   Beer's Law Using a Smartphone and Paper Device   57

9.1 Background   57

9.2 Microfluidic Device Design   59

9.3 Chemicals and Supplies   59

9.4 Hazards   60

9.5 Experimental Procedure   60

9.6 Additional Notes   62

9.7 Assessment Questions   62

References   62

10   Acid–Base Titrations on Paper   63

10.1 Background   63

10.2 Design of the Microfluidic Device   64

10.3 Chemicals and Supplies   64

10.4 Hazards   64

10.5 Experimental Procedure   64

10.6 Additional Notes   66

10.7 Assessment Questions   67

References   67

11   Simultaneous Determination of Protein and Glucose
     in Urine Sample Using a Paper-Based Bioanalytical Device   69

11.1 Background   69

11.2 Microfluidic Device Design   71

11.3 Chemicals and Supplies   71

11.4 Hazards   71

11.5 Experimental Procedure   72

11.6 Additional Notes   74

11.7 Assessment Questions   74

References   75

12   Quantitative Determination of Total Amino Acids in Tea
     Using Paper Microfluidics and a Smartphone   77

12.1 Background   77

12.2 Microfluidic Device Design   78

12.3 Chemicals and Supplies   78

12.4 Hazards   79

12.5 Experimental Procedure   79

12.6 Additional Notes 81

12.7 Assessment Questions 81

References 81

13 Determination of Nitrite Ions in Water Using Paper
Analytical Device 83

13.1 Background 83

13.2 Microfluidic Device Design 85

13.3 Chemicals and Supplies 85

13.4 Hazards 85

13.5 Experimental Procedure 85

13.6 Additional Notes 87

13.7 Assessment Questions 88

References 88

14 Colorimetric Determination of Multiple Metal
Ions on μPAD 89

14.1 Background 89

14.2 Microfluidic Device Design 92

14.3 Chemicals and Supplies 92

14.4 Hazards 93

14.5 Experimental Procedure 93

14.6 Additional Notes 94

14.7 Assessment Questions 94

References 95

15 Analysis of a Mixture of Paracetamol and 4-Aminophenol
in a Paper-Based Microfluidic Device 97

15.1 Background 97

15.2 Microfluidic Device Design 98

15.3 Chemicals and Supplies 98

15.4 Hazards 99

15.5 Experimental Procedure   99

15.6 Additional Notes   100

15.7 Assessment Question   101

References   101

16   Synthesis of Gold Nanoparticles on Microchip   103

16.1 Background   103

16.2 Microfluidic Device Design   105

16.3 Chemicals and Supplies   105

16.4 Hazards   105

16.5 Experimental Procedure   106

16.6 Additional Notes   107

16.7 Assessment Questions   107

References   107

17   Flow Synthesis of Organic Dye on Microchip   109

17.1 Background   109

17.2 Design of the Microfluidic Device   110

17.3 Chemicals and Supplies   111

17.4 Hazards   111

17.5 Experimental Procedure   111

17.6 Additional Notes   112

17.7 Assessment Questions   113

References   113

18   Protein Immobilization on a Glass Microfluidic Channel   115

18.1 Background   115

18.2 Microfluidic Device Design   117

18.3 Chemicals and Supplies   117

18.4 Hazards   117

18.5 Experimental Procedure   117

18.6 Additional Notes   119

18.7 Assessment Question   120

References   120

19 Microfluidic Enzyme-Linked Immunosorbent Assay   121

19.1 Background   121

19.2 Microfluidic Device Design   123

19.3 Chemicals and Supplies   124

19.4 Hazards   124

19.5 Experimental Procedure   124

19.6 Additional Notes   126

19.7 Assessment Questions   127

References   128

Glossary   129

Appendices   131

Index   155

About the Author

Basant Giri received BSc and MSc degrees in Chemistry from Tribhuvan University, Kathmandu, Nepal, a second MS degree in Chemistry from the Oregon State University, Corvallis, USA, and a PhD degree in Chemistry from the University of Wyoming, Laramie, USA. After working as a research fellow at Nepal Academy of Science and Technology, Nepal for six months, Dr. Giri cofounded the Kathmandu Institute of Applied Sciences in Kathmandu, Nepal. Currently he works as a scientist at the Center for Analytical Sciences at the same institute. His research interests include development of low-cost analytical devices (e.g., paper microfluidics) for biological and environmental applications. He has several years of teaching experience at high school, undergraduate, and graduate levels in Nepal and the United States as faculty and teaching assistant, respectively. Dr. Giri has authored and coauthored a textbook on Nanoscience and Nanotechnology and several peer-reviewed research articles.

Preface

Considering the increasing interest in microfluidics, *Laboratory Methods in Microfluidics* aims to fill the need for a laboratory book in this field.

Microfluidics is becoming an increasingly popular subject both in education and research. Many universities are now incorporating microfluidics in their courses to a greater or lesser extent along with experiments in the laboratory courses. Even though there are several textbooks covering this topic, there is currently no resource covering experimental procedures. This laboratory book aims to provide a number of detailed instructions for experiments in microfluidics intended for undergraduate and postgraduate courses in analytical chemistry, biochemistry, microbiology, biotechnology, environmental science, and engineering. Some experiments can even be implemented in high-school curriculum projects and experiments.

Most of the experiments described in this book have been adapted from research articles and the experience of the author while teaching undergraduate analytical chemistry labs. While care has been taken to ensure that the information in this book is correct, neither the author nor the publisher can accept responsibility for the outcome of the experimental procedures outlined in this book if not properly followed. The main aim of the book is to serve as an educational tool to prepare today's students for the more demanding regimen of microfluidics. The experiments aim to provide practical experience in the application of classical and instrumental techniques incorporated in microfluidics. Each experiment includes background information including learning objectives and an overview of the principles behind the experiment, a list of materials and chemicals required, safety notes, step-by-step procedure, additional notes to instructor, assessment questions, and recommendations for further reading. The instructions for the experiments are so detailed that the measurements can, for the most part, be taken without the help of additional literature. With *Laboratory Methods in Microfluidics* instructors no longer have to refer to many journals and books to find the right procedures for their experiments. It is assumed that students are familiar with basic laboratory techniques and procedures in science before starting experiments described in this book. However, some basic practices are covered in the Appendix.

In conclusion, this book is a work in progress, and I encourage readers to submit ideas, suggestions, and comments for improvements or for new experiments. I hope you find this laboratory manual helpful in your study.

Key features

- 18 Standalone fine-tuned experiments
- Emphasizes fabrication of microfluidic devices and their and applications
- Experiments using commonly found materials to minimize the cost
- Assessment questions for each experiment
- Appropriate illustrations for each experiment
- Additional notes for instructors allowing them to customize the experiments
- Useful information about preparation of laboratory reagents in appendices

Basant Giri
January, 2017

Acknowledgments

I am thankful to Dr. Harish Subedi of Western Nebraska Community College, Nebraska, Dr. Basu Panthi of Trinity University, Texas, and Dr. Lekh Adhikari of Rappahannock Community College, Virginia for providing input on the initial draft of this book. Likewise, I am thankful to Dr. Susma Giri (my wife), Mr. Ankit Pandeya, and Mr. Sagar Rayamajhi of Kathmandu Institute of Applied Sciences, Nepal for proofreading the manuscript.

I am grateful to my PhD advisor Dr. Debashis Dutta from the University of Wyoming, who introduced me to the field of microfluidics. Dr. Tristan Kinde of Sinclair Oil Corporation (then graduate student at Dutta group) helped me fabricate the glass microfluidic device during my early days as a PhD student. The lab methods described in this book such as fabrication of glass microfluidic device, enzyme assay, and microfluidic separation were initially developed for an instrumental analysis course by Dutta Lab at University of Wyoming.

I express my love and gratitude to my father Krishna and mother Dwarika for their love, support, patience, and sacrifice.

1

Introduction to Microfluidics

1.1 Background

The field of microfluidics has been gaining popularity in the scientific community since its jumpstart about three decades ago. This multidisciplinary field has become a unique platform for chemistry, physics, biology, materials science, fluid mechanics, and engineering disciplines in terms of understanding both fundamentals and applications. Two other terms related to microfluidics are *lab-on-a-chip* and micrototal analysis systems, popularly known as *μ-TAS*. It is important to learn microfluidic experiments considering their potential in analytical applications and their advantages over conventional analytical systems. Incorporating microfluidics in teaching laboratories enables learning opportunities for undergraduate and graduate students, even high school students and independent researchers. As microfluidics require less amount of chemicals/reagents and generate less waste, universities could reduce the cost related to chemicals and waste disposal.

1.1.1 What is Microfluidics?

Microfluidics is the science that deals with the precise control and manipulation of small volumes of fluids in network of microchannels. Generally, micro means one of the following features: small volumes (μL, nL, pL, fL) and small size leading to low energy consumption and special microdomain effects. Small size means at least one dimension of the channel must be in the range of micrometers. The behavior of fluids at microscale can differ from macroscale behavior. Factors such as surface tension, energy dissipation, and fluidic resistance start to dominate the system at micro level. A microfluidic chip or device contains a network of microchannels, which are connected to the outside of the channel by inputs and outputs pierced through the chip. Such connections serve as an interface between the macro- and microworld. Through these holes, the liquids or gas are injected and removed from the microfluidic chip. The small size of microfluidic devices offers several advantages including less sample and reagent consumption, low cost, short analysis time, portability, etc.[1]

Microfluidic technologies are not just for education and research. These technologies have now been incorporated into many commercial products. Inkjet printheads are an example of the most successful application of microfluidics.[2] Printers used to reproduce digital images produced by computers commonly use such inkjet printers. Other commercial products based on microfluidics include:

1. **Agilent bioanalyzer:** The bioanalyzer instrument provides platform for bioassays, based on electrophoresis and flow cytometry, of DNA, RNA, proteins and cells with less than four microliters of sample.[3]

Laboratory Methods in Microfluidics. DOI: http://dx.doi.org/10.1016/B978-0-12-813235-7.00001-5

2. HPLC-Chip/MS system: Produced by Agilent this system is based on microfluidic chip technology and is designed for nanospray liquid chromatography/mass spectrometry (LC/MS). According to the manufacturer, this system is robust, reliable, sensitive, and easy to use for biomarker discovery and validation, monoclonal antibody characterization, small-molecule analysis, phosphopeptide analysis, etc.[4]
3. Caliper LabChip platforms: Caliper of the PerkinElmer company has produced a number of LabChip devices/kits[5] based on microfluidics involving both electrokinetic and pressure-driven flows. These devices can be used in bioassays for drug discovery applications such as small-molecule screening, fragment based screening, target specificity profiling, etc. The genomic DNA LabChip is used for DNA analysis.[6]
4. Point-of-care blood analyzers and other medical diagnostic platforms from companies including[7] Abaxis, Abbott, Achira labs, Biosite, Biovitesse, Biolithic, Baebies, Boston Microfluidics, CardioMEMS, GenePOC, FluidMedix, Micro2Gen, Nanosphere, Nanomix, etc.

1.2 Frequently Used Microfluidic Terms

1.2.1 Laminar Flow

One of the important properties of fluid flow in the microdomain is laminar flow. In laminar or streamline flow, fluids flow side-by-side in parallel layers and do not necessarily mix unlike in turbulent type flow. Adjacent layers slide past one another like playing cards. The only mixing in laminar flow is through diffusion. Laminar flow is characterized by high-momentum diffusion and low-momentum convection. In nonscientific terms, laminar flow is *smooth* while turbulent flow is *rough*. Flow of water in steep river is an example of turbulent flow. Type of flow is characterized by a dimensionless parameter known as the Reynolds number (R_e). In microfluidic systems, the R_e is usually less than 100 and the flow is considered as laminar flow[8] (Fig. 1.1).

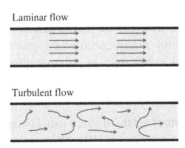

FIGURE 1.1 Schematics of laminar flow and turbulent flow.

1.2.2 Electroosmosis and Electrophoresis

These two phenomena occur when electric field is applied across the terminals of a microchannel containing fluid.

When electric field is applied across the two terminals of a microchannel, the bulk of the liquid inside the channel moves from one pole to the other. This motion of fluid is called *electroosmotic flow* (EOF), synonymously called *electroosmosis* (*see* Fig. 1.2). The velocity of this fluid flow depends on the applied voltage, microchannel material, and nature of the fluid itself. When polar liquid such as water is brought into contact into the surface of the microchannel, the microchannel surface acquires an electric charge with a thin layer of charges very close to the surface, which is known as *electric double layer*. When an electric field is applied to the fluid via electrodes placed at inlets and outlets, the net charge in the double layer is induced to move by the resulting Coulomb force. In a negatively charged microchannel surface, the EOF is directed toward the negatively charged cathode through the microchannel.[8]

The EOF velocity (ν_{EOF}) in cm/s depends on applied electric field (E) and can be calculated using Eq. (1.1). Electric field in V/cm is the applied voltage per unit length of the channel.

$$\nu_{EOF} = E \cdot \mu_{EOF} \tag{1.1}$$

where μ_{EOF} is the electroosmotic mobility, which depends on the device material and buffer solution and is defined by:

$$\mu_{EOF} = \frac{\epsilon\zeta}{\eta} \tag{1.2}$$

where ζ is the zeta potential of the channel wall, ϵ is the relative permittivity of the buffer solution, and η is the viscosity of the fluid. Experimentally, ν_{EOF} can be determined by measuring the retention time of a neutral analyte in a channel (*see* Chapter 5: Determination of

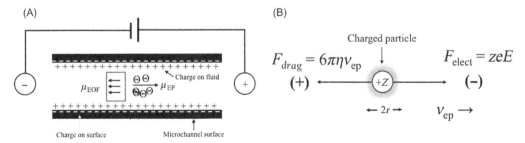

FIGURE 1.2 (A) A schematic describing EOF and EP. When electric field is applied across the two terminals of a channel, negatively charged analytes migrate toward anode with EP and bulk fluid moving toward cathode with EOF. The net flow velocity of the analyte is a compromise between EOF and EP. (B) Various forces acting upon a charged particle under the influence of applied electric field.

Electroosmotic Flow in Glass Microfluidic Device Using a Neutral Marker). The charge density of the microchannel surface depends on the pH of fluid in the channel. Therefore the EOF also depends on the pH of the fluid.

The second term, *electrophoresis*, is the motion of dispersed charged particles relative to a fluid under the influence of a spatially uniform electric field. When electric field is applied on a charged particle, the net force on the particle is the combination of electric force and dragging force from the solution as shown in Figure 1.2b.[8]

The electrophoretic (EP) velocity of the charged particle depends on the applied electric field as given by:

$$\nu_{ep} = E \cdot \mu_{ep} \tag{1.3}$$

where E is the electric field strength and μ_{ep} is the EP mobility of the particle. Polish physicist Marian Smoluchowski in 1903 developed the most well-known and widely used theory of electrophoresis. The EP mobility is given by:

$$\mu_{ep} = \frac{\varepsilon_r \varepsilon \zeta}{\eta} \tag{1.4}$$

where ε_r is the dielectric constant of the dispersion medium, ε_0 is the permittivity of free space ($C^2\,N^{-1}\,m^{-2}$), η is the dynamic viscosity of the dispersion medium (Pa s), and ζ is the zeta potential. The zeta potential refers to the electrokinetic potential of the slipping plane in the double layer. The EP mobility can be determined experimentally from the migration time and field strength:

$$\mu_{ep} = \left(\frac{L}{t_r}\right)\frac{1}{E} \tag{1.5}$$

where L is the distance from the inlet to the detection point and t_r is the time required for the analyte to reach the detection point (migration time).

The overall velocity (ν) of an analyte in an electric field, however, is a combination of both electroosmosis and electrophoresis. It can be calculated using:

$$\nu = \nu_{ep} + \nu_{EOF} = (\mu_{ep} + \mu_{EOF})E \tag{1.6}$$

Since the EOF of the bulk fluid (e.g., buffer solution) is generally greater than that of the EP mobility of the analytes, all analytes are carried along with the buffer solution toward the cathode. Depending on the charge and size, the molecules/ions move through the microchannel at different speeds, which in turn allows us to separate the molecules/ions.

Cations elute first in an order corresponding to their EP mobilities. Smaller and highly charged cations elute before the larger ones with lower charge. Furthermore, neutral species elute as a single band with an elution rate equal to the electroosmotic flow velocity.

Anions are the last components to elute, with smaller, highly charged anions having the longest elution time:

$$\nu_{cation} > \nu_{EOF}$$
$$\nu_{neutral} = \nu_{EOF}$$
$$\nu_{anion} < \nu_{EOF}$$

The electrokinetic flow (both EOF and EP) is used in a number of analytical techniques including separation. Analytes are separated based on their differing EP mobilities.

1.2.3 Separation Resolution

A generic electropherogram, which is the plot of signal obtained from separation experiment using electrophoresis, is shown in Figure 1.3. The separation resolution (R_s) between two peaks is the measure of the ability to distinguish the neighboring peaks. R_s is expressed as the ratio of the distance between the peaks (Δt) to the sum of their half widths at the base (W_{av}) and is given by:

$$R_s = \frac{\Delta t}{W_{av}} \tag{1.7}$$

If the peaks follow a Gaussian profile, the half-width of each peak can be taken as twice its standard deviation σ, yielding the following expression. By convention, a value of $R \geq 0.5$ is often taken as a criterion to indicate that two neighboring peaks are clearly distinguishable:

$$R = \frac{(t_2 - t_1)}{2(\sigma_1 + \sigma_2)} \tag{1.8}$$

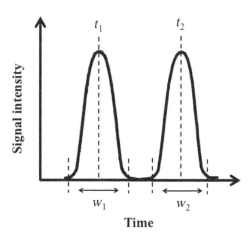

FIGURE 1.3 A typical electropherogram. The intensity of migrating analytes is recorded as they travel past a fixed detection point in the separation channel.

Treating the peaks as Gaussian, the theoretical plate number can be calculated by:

$$N = 5.54 \left(\frac{t_m}{W_{0.5}} \right)^2 \tag{1.9}$$

where t_m is the migration time and $W_{0.5}$ is the full-width at half-maximum.

In capillary electrophoresis Eq. (1.10) is used to calculate the resolution:

$$R_s = \frac{1}{4} \left(\frac{\Delta \mu_p \sqrt{N}}{\mu_p + \mu_o} \right) \tag{1.10}$$

According to this equation, the maximum resolution is reached when the EP and electroosmotic mobilities are similar in magnitude and opposite in sign. In addition, it can be seen that high resolution requires lower velocity and, correspondingly, increased analysis time. The very high resolution of electrophoresis is a consequence of its extremely high efficiency.

The number of theoretical plates or separation efficiency is given by:

$$N = \frac{\mu V}{2 D_m} \tag{1.11}$$

where N is the number of theoretical plates, μ is the apparent mobility in the separation medium, and D_m is the diffusion coefficient of the analyte. According to this equation, the efficiency of separation is only limited by diffusion and is proportional to the strength of the electric field.

The EP separation is considered better in comparison with gas and liquid chromatographic separations in terms of both separation efficiency and resolution. The efficiency of EP separations is typically much higher than the efficiency of high pressure liquid chromatography (HPLC). Unlike HPLC, in electrophoresis there is no mass transfer between phases. In addition, the EOF-driven separation does not significantly contribute to band broadening as in pressure-driven chromatography. EP separations can have several hundred thousand theoretical plates. The only contributor to peak broadening, which is responsible for poor resolution, practically, is diffusion.[9,10]

1.3 Assessment Questions

1. What are "lab-on-a-chip" devices? Give at least three advantages of such devices.
2. What are electrophoresis and electroosmosis?
3. What are the advantages of EP separation when compared to chromatographic separation?
4. How can you reduce the electroosmotic flow in a microchannel?

5. Three compounds, A, B, and C, were separated using capillary electrophoresis in a 30-m-long capillary. Among these three compounds C is a neutral molecule. The following data were obtained from the experiment:

Compound	t (min)	w (min)
A	3.1	0.16
B	3.4	0.18
C	2.3	1.01

   a. Calculate the resolution for each pair of adjacent peaks.
   b. Calculate the number of theoretical plates for each compound.

References

1. Nguyen NT, Wereley ST. *Fundamentals and Applications of Microfluidics*. Boston, USA: Artech House; 2002.

2. Volpatti LR, Yetisen AK. Commercialization of microfluidic devices. *Trends in Biotechnology*. 2014;32(7): 347−350.

3. Bioanalyzer Instruments: Agilent Technologies. Available from: <http://www.genomics.agilent.com/en/Bioanalyzer-System/2100-Bioanalyzer Instruments/?cid=AG-PT-106>.

4. Agilent 1260 Infinity HPLC-Chip/MS System: Agilent Technologies; 2016. Available from: <https://www.agilent.com/cs/library/brochures/5990-6221EN.pdf>.

5. LabChip® EZ reader mobility shift assay: PerkinElmer Company; 2016. Available from: <http://www.perkinelmer.com/lab-solutions/resources/docs/LabChip EZ Reader Assay Development Guide v2.pdf>.

6. LabChip® GX-GXII User guide: Perkin-Elmer Company; 2016. Available from: <http://www.bioneer.co.kr/literatures/manual/instrument/LabChip GX Genomic DNA User Guide.pdf>.

7. List of microfluidics companies: FluidicMEMS; 2016. Available from: <http://fluidicmems.com/list-of-microfluidics-lab-on-a-chip-and-biomems-companies/>.

8. Kirby BJ. *Micro- and Nanoscale Fluid Mechanics: Transport in Microfluidic Devices*. New York: Cambridge University Press; 2010.

9. Landers JP, ed. *Handbook of Capillary and Microchip Electrophoresis and Associated Microtechniques*. 3rd ed. New York: CRC Press; 2008.

10. Harris DC. *Quantitative Chemical Analysis*. 8th ed. New York: W.H. Freeman and Company; 2010.

2

Fabrication of a Glass Microfluidic Device

2.1 Background

Glass, the most commonly found glassware in chemistry laboratories, is one of the oldest materials used in microfluidics and is still widely used for research and commercial products. There are basically three types of glasses used in microfluidic devices: soda lime, quartz, and borosilicate. Glass is primarily made up of silicon dioxide (SiO_2) mixed with other oxides (B_2O_3, Na_2O, CaO, MgO, etc.) in different proportions. The properties[1] of these three types of glass are presented in Table 2.1.

Glass has many advantages over other microfluidic materials. Its properties make it a very suitable material for microfluidic applications such as chemical reactions, chromatography, fluorescence detection and optical applications, capillary electrophoresis (CE), polymerase chain reactions (PCR), gas chromatography, etc. The advantages[2] of using glass devices include:

1. Because of its excellent chemical resistance to many chemicals, experiments utilizing organic solvents and acidic or alkali reagents are typically suitable with a glass chip.
2. Surface-modification procedures of glass are well established making the covalent coupling of biochemical in immunoassays easier.
3. Optical transparency and low background fluorescence combined with low auto-fluorescence makes glass suitable for optical detection (e.g., chemiluminescence, fluorescence measurement) system.
4. The biocompatibility of glass makes it suitable for applications that utilize biological reagents.
5. Glass has high resistance to mechanical stress.
6. Because of its high-temperature resistance, thermal stability, and low thermal expansion coefficients, applications demanding high-temperature resilience (e.g., chemical synthesis) can be carried out in glass devices.
7. Glass is also good for electrophoretic-based separation applications.
8. Glass is hydrophilic and doesn't allow diffusion of small molecules from channel.

   A photograph of a microfluidic device made from borosilicate glass is shown in Fig. 2.1.

There are many ways to make glass microfluidic devices but generally fabrication involves combination of photolithography, wet chemical etching, and bonding (Fig. 2.2).

Laboratory Methods in Microfluidics. DOI: http://dx.doi.org/10.1016/B978-0-12-813235-7.00002-7

Table 2.1   Properties of Different Types of Glass

Properties	Soda Lime	Borosilicate	Quartz
Chemical composition (wt%)	$SiO_2 = 74$, $Na_2O = 13$ $CaO = 10.5$, $Al_2O_3 = 1.3$ $K_2O = 0.3$, $SO_3 = 0.2$ $MgO = 0.01$, $TiO_2 = 0.01$ $Fe_2O_3 = 0.04$	$SiO_2 = 81$, $B_2O_3 = 12.5$ $Na_2O = 4$, $Al_2O_3 = 2.3$ $CaO = 0.02$, $K_2O = 0.06$	$SiO_2 = 100$
Viscosity $(\log(\eta)$, Pa s$) = A + B/(T$ in $°C − T_o)$	550–1450°C: $A = -2.309$, $B = 3922$ $T_o = 291$	550–1450°C: $A = -2.834$, $B = 6668$ $T_o = 108$	$\log n = 12.7$
Transition temperature (T_g) [°C]	573	620	1200
Thermal expansion coefficient (K) $\sim 100-300°C$ [10^{-6}/°C]	9	3.5	5.5
Density at 20°C (kg/m^3)	2.5×10^3	2.2×10^3	2.2×10^3
Refractive index (n_D) at 20°C	1.52	1.47	1.45
Young's modulus, elasticity (E) at 20°C (N/m^2)	0.72	0.65	0.72
Shear modulus, rigidity (G) at 20°C (GPa)	28.2	25	31
Annealing temperature (°C)	546	560	1140
Specific heat capacity at 20°C [J/(mol K)]	49	50	45.3

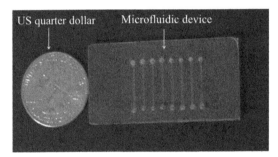

FIGURE 2.1 Image of glass microfluidic device. Comparing the size with US quarter dollar.

FIGURE 2.2 Three major steps for the fabrication of glass microchips.

2.1.1 Photolithographic Procedure

Photolithography[3] is a process of transferring channel design created on a photomask to a sacrificial thin film called photoresist coated on glass substrate. It simply sketches the outline of the future device on photoresist.

The first step of the microfluidic device fabrication process involves production of a photomask with a layout of the architecture of the channel network design. The photomask is used to transfer the channel patterns to a photoresist-coated glass substrate. The microfluidic channel networks are typically designed using drawing software like CADopia, Adobe Illustrator, AutoCAD, L-edit, CleWin, Cadence, Layout Editor, Corel-DRAW, etc. The network is then printed onto transparency film or chromium using high-resolution printers. The design can be printed on photomask from a commercial printing company (e.g., Digidat, Fineline Imaging, etc.).

Before performing photolithography, we have to have the substrate ready. The glass substrate is typically coated with an adhesion layer (e.g., chromium) and a photoresist, each of which is hundreds of nanometers thick, on only one side of substrate. The chromium layer reduces the etch undercut and improves the quality of the channels produced. Photoresist is a thin layer of polymerized organic material deposited on top of the chromium layer on the glass substrate. The resist comprises two parts: an active polymer component sensitive to UV radiation and a matrix component insensitive to the radiation but fulfilling the mechanical requirements of the resist. There are two types of resists: positive and negative. Upon exposure to the radiation, a positive resist becomes soluble whereas the negative resist becomes insoluble after soaking in developing solution. The resist layer is locally modified when selectively exposed to UV radiation using lithography tools.[3] A positive photoresist thus forms positive images of the mask patterns on the substrate.

These adhesive layers can be coated in the lab, for example, by a spin-coating procedure if the facility is available. If not, glass substrate coated with these two layers can be purchased from a commercial vender (*see* Appendix VI).

Both the substrate and mask should be clean without any airborne debris or dust during photolithography procedures. The imbedded particle could result in defects on the glass substrate and reduce resolution. Pure N_2 can be used to remove surface particles from the photomask and substrate. As the distance between the photomask and substrate determines the resolution, uniform contact between the photomask and substrate is critical for reproducible patterning. Both the mask and substrate must have perfectly planar surfaces. The narrowest feature size that can be patterned using photolithography is limited by diffraction to around 0.5 μm for contact photolithography. When even smaller features are required, electron-beam or nanoimprint lithography may be used instead.[3]

During UV exposure, polymer of photoresist breaks down and is removed from the areas exposed to the radiation. The microchannel pattern on the glass substrate after this step is clearly seen. The duration of exposure depends on the exposure dose required for the photoresist and on the UV lamp power. We can calculate[4] the exposure times using:

$$\text{Exposure time (seconds)} = \frac{\text{Exposure done (mJ/cm}^2)}{\text{Effective power (mW/cm}^2)} \qquad (2.1)$$

The photolithographic step is followed by chemical wet etching to obtain the desired channel depth using glass etching followed by sealing channels with a cover plate through the bonding process. Before bonding, access holes at the end of channels are made using a microabrasive power-blasting system.

2.1.2 Chemical Wet Etching

Channels on the device are made using a chemical wet-etching method. In wet etching, the sample surface is etched chemically using a chemical that is specific to the substrate but inert to the resist layer. The wet-etching method is very simple and easy to implement. The buffer-oxide etchant (BOE)[5] solution that contains a variable amount of hydrofluoric acid (HF) will be used for glass etching in this experiment. The BOE solution can be prepared in the lab or can be purchased from a commercial vendor (e.g., Transense, Inc., USA). Transense BOE (1:10) contains a mixture of HF (1%–25%), ammonium fluoride (NH$_4$F) (2%–40%), and distilled water. In some cases NH$_4$F is replaced with HNO$_3$.

The main disadvantage of wet etching of glass with HF is the isotropic etching,[3] i.e., the glass surface is etched in all directions within the pattern resulting in a low aspect ratio and considerable broadening of the pattern after etching. Therefore the lateral dimensions of the structures are no longer precisely controlled. Isotropic etching produces channels with a width approximately twice the depth and only semielliptical bro-section channels are possible. Isotropic etching inherently has rounded corners at the base of the channel with a radius roughly equal to the channel depth (*see* Fig. 2.3). Consequently, it is not possible to fabricate channels with sharp corners through the application of this method.

Channel width (W_C) in this case can be estimated from the mask width (W_M) and the channel depth (D_C), which are related by:[3]

$$W_C = 2D_C + W_M \tag{2.2}$$

2.1.3 Drilling Access Holes

Once desired channel depth is obtained on the substrate, access holes are punched at the end of each channel by drilling or power blasting. Access holes are used to provide liquid access to the channels in microfluidic chips. Since the device is small and fragile, care must

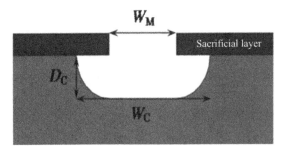

FIGURE 2.3 A cartoon showing side view of isotropic etching.

be taken when handling or drilling apertures so that the glass is not broken or the port does not come into contact with more than one channel.

Access holes are drilled on the substrate plate or cover plate. When drilled on the substrate, the drilling should be from the surface that is opposite to the etched channels. One of the simplest methods to make access holes is by using a microabrasive power-blasting system. In this system, pressurized sand particles are blown from the backside of the substrate. The microabrasive system utilizes mechanical erosion on the substrate by bombarding it with high-kinetic energy sand particles, which results in conically shaped holes (~ 1 mm). The size of the holes can be controlled by varying the size of nozzle tips and sand particles. After drilling access holes, the etched chip is aggressively rinsed with water to remove all glass fragments and sand particles. Acetone and chrome etchant are then used to remove the remaining photoresist and chrome from the etched substrate before starting the bonding procedure.

2.1.4 Bonding Two Glass Plates

This step involves the sealing of channels on the substrate with a cover plate to prevent leakage and ensure the flow can be manipulated through the desired channels. The bonding is critical for chemical and biological applications to make sure there is no contamination of reagents and the device withstands the internal pressure arising from the fluid flow in the small channels.

A number of glass-glass bonding techniques are described in the literature and choosing an appropriate bonding process is a critical step in the microfluidic chip fabrication process. In this experiment, we will follow a room temperature (RT) direct bonding[6] method of the two glass plates. This bonding method is simple and does not require clean room facilities, programmed high-temperature furnaces, pressurized water sources, and adhesives. The bonding process can be completed in less than 3 hours. More than 95% high bonding yield can be achieved using the RT bonding method.

Ensuring a clean surface of substrate and cover plate is probably the most important requirement for RT glass bonding in addition to the surface flatness. Therefore rigorous cleaning is necessary. The most unsuccessful bonding events are mostly associated with solid particles or organic material remaining on the glass surfaces, for example, dust, residual photoresist or glass particles, oil from fingertips, etc. There are several methods[6] for cleaning glass plates and generally a combination of these methods is required to ensure a very clean glass surface.

Ablation: Ablation is used to remove material by microetching mainly to increase surface area. It is also used to remove a weak boundary layer.

Plasma ashing: Plasma ashing is a process of removing photoresist from an etched substrate. A plasma source created monoatomic reactive species (e.g., oxygen or fluorine) combine with the photoresist to form ash, which is removed with a vacuum pump. It removes the surface contamination and prevents any contamination from interfering with the bonding process. It also makes the surface more hydrophilic.

Sulfuric acid: Concentrated sulfuric acid is used to clean the glass surface by soaking the glass plates for several hours at RT. The soaking of glass plates in sulfuric acid has proved to significantly contribute to the formation of a hydrolyzed layer on the glass surface, which is important for improving bonding quality and yield at RT.

Piranha solution: Piranha is a mixture of sulfuric acid and hydrogen peroxide and is the most commonly used solution to clean glass surface. As piranha solution contains strong oxidizing agent, it is used to aggressively clean organic materials.

Organic solvents: Water-soluble organic solvents (e.g., acetone, ethanol) are also used to clean glass surface.

Water: A good method to aggressively rinse the etched chip is to pinch a hose attached to a water spigot. High-flow-rate tap water has shown better effects for removing minute solid particles from chip surfaces. To further facilitate removal of glass particles, the etched glass is placed (channels down) in a clean beaker and sonicated for 10–15 minutes in water.

The time interval for which the two glass plates are kept together during bonding is important. It has been reported that the bonding strength increases with standing time and reaches a maximum after some time.[6] Surface flatness of the glass plates is another factor that affects close contact of surfaces and therefore bonding quality. The bonding yield is strongly dependent on the chip area.

When glass plates are treated with water, a hydrolyzed gel layer is formed. Most of the Si–ONa groups near the surface of the chips are transformed into Si–OH groups. The two hydrolyzed glass-plate surfaces are initially held together by hydrogen bonding and then joined by siloxane bond formation. The Si–OH groups gradually get dehydrated forming siloxane bonds and terminating with a condensation–polymerization reaction forming Si–O–Si bonds between the two pieces of glass. The natural dehydration process is slow and apparently is determined by the speed of water evaporation from the bonded surface.[6]

2.1.5 Putting Fluid Reservoirs

While some applications may not require external reservoirs in glass chips, many applications require introducing and recovering fluids (e.g., samples, reagents, or buffers). Generally a fluid reservoir is used for the storage and distribution of sample or buffers or reagents.

Some commonly used reservoirs are described as follows:

Pipette tips: Cut polypropylene pipette tips are inserted and glued into each access hole with epoxy or UV glue to serve as solution reservoirs.

Glass reservoirs: Cylindrical glass reservoirs can be affixed on the access holes of the microfluidic chips using glue to serve as reservoirs. Hollow Pyrex glass cylinders are the cheapest and most commonly used reservoirs. The height and inner diameter of such glass reservoirs can be chosen based on need. Such reservoirs have a larger opening that allows for the quick addition or removal of fluids as well as easier application of electric field via a platinum wire.

Commercial flow adapters: Microfluidic flow adapters are commercially available from companies like Upchurch, Swagelok, and Valco. These commercial fittings are more applicable when one is required to connect the chip to other analytical instrumentations like mass spectrometry or to syringe pumps.

In this experiment, students will fabricate glass microchip combining all three major steps outlined above. This laboratory exercise is intended for a 4-hour lab period in a senior-undergraduate and graduate-level course in chemistry or engineering. After the completion of this experiment, students will be familiar with the microfabrication procedures including photolithography, wet etching, and RT bonding.

2.2 Microfluidic Device Design

A straight-channel design (*see* Fig. 2.4) is suggested for simplicity. The chip is 2″ × 1″ and has seven straight channels. Each channel is 15 mm long and 0.5 mm wide. Every channel has access holes at both ends. These layout designs of the chip can be made using Microsoft PowerPoint or other advanced drawing software such as CADopia, Adobe Illustrator, etc. It is then printed onto transparency films or chromium using high-resolution printers. Commercial printing company can be used (e.g., Digidat, Fineline Imaging, etc.) for getting the photomask.

2.3 Chemicals and Supplies

Deionized (DI) water, developing solution, chromium etchant, buffered oxide etchant (BOE), nitrogen gas, photoresist, and fluorescent dye solution (e.g., rhodamine B).

Glass substrate coated with chromium and photoresist, cover-plate glass, UV lamp, photomask with desired design, photolithography box or holder, oven, glass cutter, hot plate with stirrer, XY profiler, plastic tweezer, crocodile clips, Teflon jars, magnetic rod, sandblaster, beaker, and paper towel.

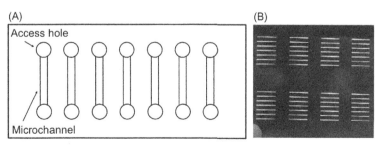

FIGURE 2.4 (A) Schematic of straight-channel microfluidic design, and (B) photograph of a photomask as example. Eight microchips can be made from this mask.

2.4 Hazards

Always wear safety goggles and a lab coat when working in the laboratory. Special care must be taken when handling HF solutions because of the chemical's ability to permeate skin insidiously and dissolve bone. The chemical etching must be carried out under fume hood.

2.5 Experimental Procedure

2.5.1 Photolithography

Photography can be performed with a group of 4–8 students. The photolithography box or holder can be made in the lab. This box is supposed to house the substrate plate, photomask, and another glass plate. It is better to have a cover to protect from light on it (*see* Fig. 2.5).

1. Turn off all light sources including computer screens in the photolithography room and turn on the UV lamp at least 15 minutes before exposure.
2. Pour some developing solution in a large petri dish that fits the substrate to soak in.

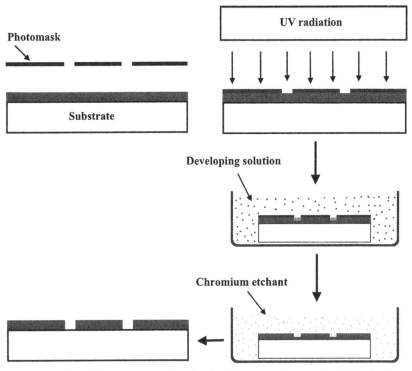

FIGURE 2.5 Schematic of photolithography procedure for the fabrication of glass microfluidic device.

3. Place the glass substrate on a holder. Then keep the photomask on top of it. Put another glass plate on the top.
4. Expose the assembly to the UV radiation for ∼40 seconds.
5. Remove the assembly and soak the substrate in the developing solution for ∼5 minutes.
6. Wash the substrate with DI water and dry it by blowing nitrogen gas.
7. You may want to put photoresist on a nonchannel area of the substrate and dry it in the oven at 80°C to protect the surface from being scratched.
8. Soak the substrate in chromium etchant for ∼20 minutes. Remove and wash with DI water. At this point, the channel network is clearly visible through the glass.

2.5.2 Chemical Wet Etching

This step (*see* Fig. 2.6) can be done individually or by a group of two students.

1. Cut the substrate into eight individual chips using a glass cutter.
2. Put photoresist on the backside of the substrate and nonchannel areas on the front side. Dry in oven at 80°C for ∼10 minutes.
3. Immerse the substrate in BOE (1:10) solution contained in a Teflon coated jar for 10 minutes. Make sure the BOE solution is stirred (e.g., 120 rpm) and heated (e.g., 60°C) using a hot plate.
4. Remove from BOE solution, wash with DI water, dry with nitrogen gas, and check the depth of the channel using an XY profiler.
5. Repeat steps 3 and 4 until the desired depth is achieved.
6. Bring the etched substrate to a sandblaster to drill access holes. Position the tip of the sandblaster nozzle right above the location of the access hole from the backside of the substrate.
7. Drill the access hole at both ends of the channel.
8. After drilling access holes, rinse the etched substrate aggressively with water to remove all glass fragments and sand particles. Dry it by blowing nitrogen gas.
9. Remove the remaining photoresist by spraying acetone.

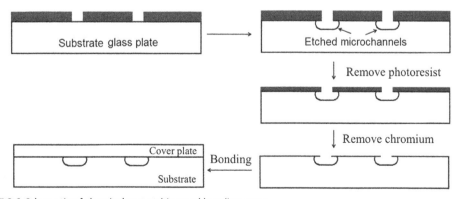

FIGURE 2.6 Schematic of chemical wet-etching and bonding steps.

10. Remove chromium from the entire substrate by soaking the substrate in chrome etchant for ~30 minutes.

2.5.3 Bonding of Two Glass Plates

We are going to do a RT bonding procedure. At this point you will have the substrate etched and drilled and a cover plate of the same size (Fig. 2.7).

1. Clean both glass plates (bottom substrate and cover plate) with detergent solution and then rinse with by running stream of water. It is best to trace the etched regions with the water stream, ensuring that no debris is left in the channels.
2. Immediately after cleaning, bring the plates together into contact in a beaker containing water.
3. Remove the plates from the beaker together using a plastic tweezer.
4. Remove water from the plates by gently pressing them with a paper towel without detaching them.
5. Keep the assembly together under some pressure using crocodile clips for ~2 hours.
6. Check the chip for interference fringe.
7. The chip without fringe is then placed in an oven (e.g., 80°C) for ~1 hour to further strengthen the bonding. If fringe is seen, dissemble the glass plates, clean, and bond again.

2.5.4 Evaluation of Chip

1. Apply gentle pressure using fingers on both sides of the chip. If you see interference fringes, the bonding is not good.
2. Once the chip passes the fringe test, put a drop of dye solution at one side of the channel-into the access hole and allow it to flow in the channel.
3. Observe the flow of dye solution visually with an eye or with a fluorescent microscope. The spread of dye solution outside the channel area can be considered to be due to a leaking chip (Fig. 2.8).

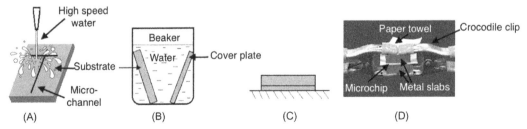

FIGURE 2.7 Schematic of room-temperature bonding procedure. (A) High-speed water stream is used to clean the glass plates; (B) two glass plates are brought in contact in a beaker containing clean water; (C) the glass plates are placed on top of each other; (D) a crocodile clip is used to apply some pressure on glass plates. *Partly adapted with permission from Jia ZJ, Fang Q, Fang ZL. Bonding of glass microfluidic chips at room temperatures. Anal Chem. 2004;76(18):5597–5602. Copyright 2004 American Chemical Society.*

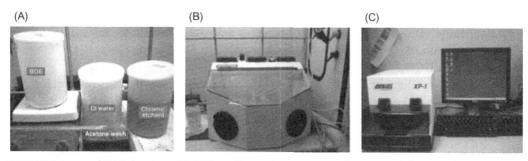

FIGURE 2.8 Image of (A) Teflon jars with BOE, washing water, acetone wash waste, and chrome etchant (B) sandblaster, and (C) XY profiler.

2.5.5 Lab Report

Write a lab report following ACS style for journal articles. It should include an abstract, introduction, experimental, results and discussion, and conclusion sections with diagrams, tables, calculations, and images wherever necessary.

Consider following points when writing the report:

1. Include a short description of fabrication of a glass microfluidic device.
2. Outline the major steps and workflow involved in the experiment. Be sure to report the experimental conditions for the assay and the operational parameters for the instrumentation used.
3. Include representative photographs of device and experimental settings.
4. Add the results of device evaluation test.

2.6 Additional Notes

Consider the following notes that may be useful for successful fabrication of a glass microchip:

1. A very simple design is suggested in this experiment. However, instructors can suggest alternate designs depending on the requirement.
2. If you do not have a facility to print the photomasks, it can be bought from commercial vendors and should be provided to students.
3. This experiment does not incorporate the coating procedure. Therefore it is suggested to buy precoated glass substrate with photoresist and chromium layers. It can also be coated in the lab if the facility is available and time permits.
4. In the case of underexposure during UV irradiation, the chemical reaction will not be completed leading to incomplete solubility of the photoresist thus creating a partially developed pattern. Overexposure of photoresist could spread the chemical reaction outside the bounds of the exposed area resulting in ragged edges and ballooned patterns compromising channel size and resolution.
5. It is important to note that leaving the substrate plate in chromium etchant solution for a long time can remove photoresist and chromium layers from the entire plate.

6. The BOE solution can be prepared in the lab or purchased from a commercial vendor (*see* Appendix VI). Transcense BOE (1:10) contains a mixture of HF (1%–25%), ammonium fluoride (NH_4F) (2%–40%), and distilled water. In some cases, NH_4F is replaced with HNO_3.

7. The insoluble products generated during the etching process may be deposited on the etched surfaces and act as masking layers leading to roughening of the surface of the etched channel. Continuous stirring of the BOE solution can minimize this deposition. The etch rate depends on HF concentration, glass composition, temperature, and extent of the agitation, and can be manipulated. Etch rate can be increased by increasing the temperature of the solution and by increasing the concentration of HF.[7]

8. Unsuccessful bonding events are mostly associated with solid particles or organic material remaining on the glass surfaces. For example: dust, residual photoresist, or glass particles, oil from fingertips, etc.[6] Concentrated sulfuric acid is used to clean the glass surface by soaking the glass plates for several hours at RT. A piranha solution is also used for aggressively cleaning organic materials. Water-soluble organic solvents (e.g., acetone, ethanol) are also used to clean glass surface. You can use a stack of books to provide bonding pressure.

2.7 Assessment Questions

1. Calculate the channel width (W_C) in a chip that was created using the mask width (W_M) of 500 µm and the channel depth (D_C) of 30 µm.
2. Write down the reaction of glass etching with BOE.
3. Point out five important factors responsible for successful RT bonding of a glass microfluidic device.

References

1. Properties of glass: Department of chemistry and biochemistry, University of Delaware. Available from: <http://www1.udel.edu/chem/GlassShop/PhysicalProperties.htm>.

2. Iliescu C, Taylor H, Avram M, Miao J, Franssila S. A practical guide for the fabrication of microfluidic devices using glass and silicon. *Biomicrofluidics*. 2012;6(1):16505–1650516.

3. Mailly D, Christophe V. *Lithography and Etching Processes in Nanoscience*. Berlin: Springer; 2007.

4. Ferry MS, Razinkov IA, Hasty J. Microfluidics for synthetic biology from design to execution. In: Voig C, ed. *Methods Enzymol*. 497, 2011:344.

5. Williams KR, Muller RS. Etch rates for micromachining processing. *J Microelectromech Syst*. 1996; 5(4):256–269.

6. Jia ZJ, Fang Q, Fang ZL. Bonding of glass microfluidic chips at room temperatures. *Anal Chem*. 2004; 76(18):5597–5602.

7. Castaño-Àlvarez M, Pozo Ayuso DF, Granda MG, Fernández-Abedul MT, García JR, Costa-García A. Critical points in the fabrication of microfluidic devices on glass substrates. *Sens Actuators B*. 2008;130:436–448.

Fabrication of a Paper Microfluidic Device for Blood-Plasma Separation

3.1 Background

Blood tests are widely used in clinical diagnosis for monitoring health status. Blood is a complex mixture of bodily fluid containing about 55% plasma and the rest blood cells primarily red blood cells (RBCs). About 92% of plasma contains water and the remaining 8% is dissolved proteins, minerals, glucose, etc.[1] Most of the colorimetric assays used in clinical tests need purified blood plasma for analysis. Blood-plasma separation is a particularly important step for colorimetric assays as the intense red color of RBCs may interfere with quantification. As a result, biochemical tests are typically carried out in serum or plasma, necessitating a separation method such as centrifugation or sedimentation. Conventional blood-separation methods based on centrifugation or magnetic separation are effective, but are time consuming and require an additional sample-preparation step to isolate plasma from whole-blood samples taken from patients.

In this experiment[2], we will fabricate a simple microfluidic paper-based analytical device (μPAD) using a wax-dipping method for blood-plasma separation to demonstrate the separation of plasma from whole-blood samples. We will verify this separation method by measuring the presence of proteins in whole-blood samples using a colorimetric test. The reproducibility of the colorimetric measurements on the μPAD will also be studied by multiple measurements. However, before jumping into the experimental details, let's learn briefly about the μPAD.

3.1.1 Paper Microfluidics

The Whitesides group from Harvard University was the first to introduce the concept of paper microfluidics, also known as $\mu PADs$, in 2007. $\mu PADs$ consist of a patterned, hydrophilic channels on paper separated by hydrophobic barriers that direct the flow of fluids. Paper is emerging as a low-cost platform for the construction of microfluidic devices. Paper functions as 3D microfluidic substrate because its cellulose fiber network acts as capillaries, wicking aqueous solutions without the need for active pumping. Paper is an attractive substrate[3] for microfluidic devices because:

1. It is abundant and cheap.
2. It is easy to use, store, transport, dispose, can be recycled, and is biodegradable.
3. It has a high surface area for visualization.
4. It is easy to modify chemically.

Laboratory Methods in Microfluidics. DOI: http://dx.doi.org/10.1016/B978-0-12-813235-7.00003-9

Depositing a hydrophobic material such as photoresist or wax on paper creates channels on paper devices. This experiment utilizes melted wax as a hydrophobic barrier. Waxes are hydrophobic organic compounds containing long alkyl chains. These compounds are not soluble in water but are soluble in organic solvent. The melting point of wax depends on the type of functional groups and alkyl chain and is generally above 40°C. One of the most common techniques for quantifying analytes on paper is colorimetry.

Colorimetric sensors are attractive for analytical measurements because they offer a high-contrast signal that is easy to quantify with an external optical reader such as a scanner, camera, or smartphone. Both quantitative and qualitative analyses can be carried out with $\mu PADs$ using colorimetric detection. Camera and scanner, in conjunction with imaging software, have been used to measure the size and intensity of color spots on $\mu PADs$, and these portable, relatively inexpensive technologies allow for electronic transmission of data from the field to the laboratory. The filter paper's high water-absorbing capacity provides a good medium for the colorimetric reaction to take place, which makes filter paper an ideal candidate for $\mu PADs$.

3.1.2 Protein Assay

Bromocresol green (BCG) colorimetric assay is a widely used method for protein detection.[4] In this assay, BCG (Fig. 3.1) binds with proteins to cause a dramatic color change from colorless to deep blue. This reaction is allowed to occur in the paper device and images of the detection zones are captured using a digital camera or smartphone. The color intensity of sample is then analyzed using ImageJ software in grayscale mode and compared with the intensity of blank.

The BCG assay method utilizes bromocresol green indicator, which forms a colored complex with protein, specifically albumin. This assay is used to directly measure the amount of protein without any pretreatment of samples, such as serum, plasma, urine, etc. The color of this indicator changes at pH (4.0) when it binds to a protein molecule. The intensity of the color is directly proportional to the albumin concentration in the sample. The normal range of serum albumin in our body is 35−50 g/L.[5] Decrease in serum albumin may occur in protein undernutrition, intestinal malabsorption, protein-losing enteropathy, liver disease, wasting diseases, nephritic syndrome, and haemodilution. A severe

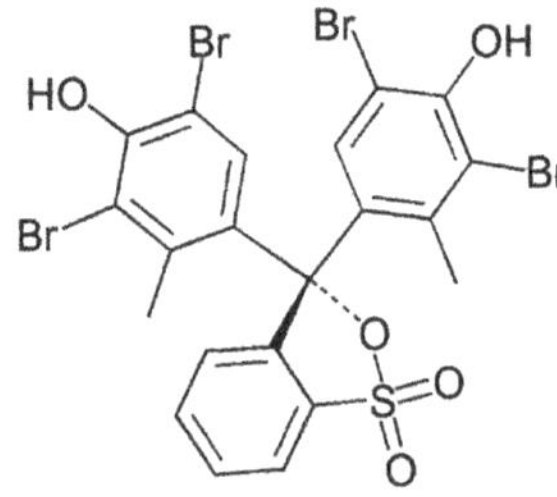

FIGURE 3.1 Structure of bromocresol green molecule.

decrease may be seen in analbuminaemia. An increase in serum albumin may occur in dehydration due to haemoconcentration. Using the BCG assay method in *μPADs* will prove to be a simple, easy-to-use, and low-cost approach in detecting the levels of albumin in blood sample.

This exercise is suitable for undergraduate labs for biochemistry or clinical biochemistry courses. After completing this experiment, students will learn the fabrication of *μPADs* by wax-dipping method, separation of blood plasma, colorimetric determination of protein, etc. This lab can be completed in a 2−3 hours lab session.

3.2 Microfluidic Device Design

The *μPAD* for this experiment consists of three circular hydrophilic zones (Fig. 3.2).[6] One for sample loading (8 mm in diameter) and the other two for detection (3 mm in diameter) of plasma in whole blood. The sample-loading zone is connected to the detection zones via microchannel (1 mm wide). The sample-loading zone and detection regions are created in different types of papers. The sample-loading zone as it is used for separation of RBCs and plasma is made from blood-separation paper whereas the detection zones are created in regular filter paper such as Whatman No.1. The separation membrane removes particles greater than 2−3 μm including red cells and platelets. Normally, RBCs are 6−8 μm in diameter and thus are trapped on the filter membrane. The device does not require external pumping forces to perform the separation. A drop of whole blood is placed onto the sample-loading zone. The plasma then wicks laterally into the test zones of the device by capillary action where protein in plasma reacts with the reagents of the colorimetric test to produce a visible color change.

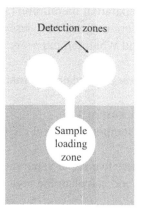

FIGURE 3.2 Design of the *μPAD* for blood-plasma separation. *(Adapted in part from Songjaroen T, Dungchai W, Chailapakul O, Henry CS, Laiwattanapaisal W. Blood separation on microfluidic paper-based analytical devices.* Lab Chip. *2012;12(18):3392−3398 with permission from the Royal Society of Chemistry.)*

3.3 Chemicals and Supplies

38% Hematocrit blood, BCG, deionized water, sodium hydroxide, and succinic acid.

Whatman No.1 filter paper, blood-separation paper LF1, iron template, white wax pellets, magnetic bar, glass slide, hot plate, digital camera or smartphone, and ImageJ software.

3.4 Hazards

Blood sample should be handled carefully. Sodium hydroxide is corrosive. BCG may be irritating and contact with skin and eyes should be avoided. Succinic acid may be irritant to skin and eyes. Hot wax should be handled carefully.

3.5 Experimental Procedure

Solutions can be prepared in group. The remaining steps are recommended to be carried out by individual students.

3.5.1 Solution Preparation

1. BCG solution: Dissolve 419 mg of BCG in 10 mL of water. Mix 250 mL of BCG solution with 750 mL of succinate buffer and adjust the pH to 4 with 0.1 N sodium hydroxide solution.
2. Succinate buffer: Dissolve 11.8 g of succinic acid in 800 mL of water and adjust the pH to 4.0 with 0.1 N sodium hydroxide.
3. Blood samples: Obtain standard blood sample from instructor. Buffer solution can be used as a blank.

3.5.2 Fabrication of the Device

1. Assemble iron template, blood-separation membrane (1 cm diameter), Whatman filter paper (2.5 cm diameter), permanent magnet, and glass slide as shown in Figure 3.3. Overlap separation membrane and Whatman paper by $\sim$1 mm. The width of channel of an iron mold is 1 mm.
2. Melt white bees wax in a beaker on a hot plate and maintain the temperature at 125°C throughout the experiment.
3. Dip the assembly into the melted wax for 1 s.
4. When the wax is cooled to room temperature, peel the paper off the glass slide and separate from the iron mold.
5. Repeat the procedure so to produce five paper separation devices.
6. Take photographs of the devices and calculate the area of detection zones.

3.5.3 Separation Procedure

1. Out of the two detection zones, one is used for protein sample and the second is used for blank. First, drop 0.5 μL of 10 × BCG working reagent on one of the

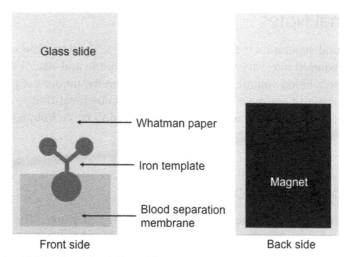

FIGURE 3.3 Assembly of fabrication units. *(Adapted from Songjaroen T, Dungchai W, Chailapakul O, Henry CS, Laiwattanapaisal W. Blood separation on microfluidic paper-based analytical devices.* Lab Chip. *2012;12 (18):3392–3398 with permission from the Royal Society of Chemistry.)*

detection zones of the μPAD. Drop the same amount of buffer solution in the black detection zone. Allow the detection zones to dry at room temperature for 10 min.

2. Add 10 μL of blood sample onto the sample-loading zone of the microfluidic device.
3. Observe the flow of plasma from the sample-loading zone to the detection zone. The color is developed after the binding of protein to BCG in the detection zone.
4. Repeat this procedure for five paper separation devices.
5. Take photographs of all five devices using your smartphone.
6. Analyze the intensity in both detection zones and measure the gray value using ImageJ (*see* Appendix V).
7. Estimate the mean signal for the blood-plasma protein level compared to blank and also calculate the standard deviation using five measurements.

3.5.4 Lab Report

Write a lab report following ACS style for journal articles. It should include an abstract, introduction, experimental, results and discussion, and conclusion sections with diagrams, tables, calculations, and images wherever necessary.

Consider the following points when writing the lab report:

1. Include a short description of wax-dipping method, blood-plasma separation, and BCG assay method for protein determination.
2. Outline the major steps and workflow involved in the experiment. Be sure to report the experimental conditions for the assay and the parameters for the camera settings used.
3. Add representative photos of device and assay regions.
4. The mean and standard deviation of signal should be obtained.

3.6 Additional Notes

1. You can use blood-separation membrane other than LF1. However, the volume of the blood sample required may vary with the type of membrane and size of the sample-loading zone used. Blood volume can also vary based on the amount of protein it has.
2. Other target analytes in plasma such as glucose can also be measured.
3. Quantitative measurement can be done using a calibration curve from a blood sample containing a known amount of the analyte.

3.7 Assessment Questions

The volume of blood required depends on the separation-paper area. The following linear equation $y = 0.431x\ 2\ 2.60$ for LF1 membrane where y is blood volume and x is paper area was obtained. Estimate the exact whole-blood volume needed to apply to your device.

References

1. Tietz NW. *Clinical Guide to Laboratory Tests.* Philadelphia: W.B. Saunders Co; 1995.

2. Songjaroen T, Dungchai W, Chailapakul O, Laiwattanapaisal W. Novel, simple and low-cost alternative method for fabrication of paper-based microfluidics by wax dipping. *Talanta.* 2011;85(5):2587−2593.

3. Li X, Ballerini DR, Shen W. A perspective on paper-based microfluidics: current status and future trends. *Biomicrofluidics.* 2012;6(1):011301−011313.

4. Sapan CV, Lundblad RL, Price NC. Colorimetric protein assay techniques. *Biotechnol Appl Biochem.* 1999;29(2):99−108.

5. Rustad P, Felding P, Franzson L, Kairisto V, Lahti A, Martensson A, et al. The Nordic Reference Interval Project 2000: recommended reference intervals for 25 common biochemical properties. *Scand J Clin Lab Investig.* 2004;64(4):271−284.

6. Songjaroen T, Dungchai W, Chailapakul O, Henry CS, Laiwattanapaisal W. Blood separation on microfluidic paper-based analytical devices. *Lab Chip.* 2012;12(18):3392−3398.

4

Fabrication and Testing of a PDMS Microchip

4.1 Background

Polydimethylsiloxane (PDMS) is a silicon-based elastomeric organic polymer. It is widely used for the fabrication of microfluidic systems because it can readily be molded into a desired shape. PDMS is easy to bond with another substrate to seal channels. This polymer substrate is transparent and therefore allows visualization of the sample inside the channel. Moreover, it is inert, nontoxic, biocompatible, inexpensive, and easy to handle.[1]

In this experiment, we are going to fabricate a microfluidic device using PDMS. We will also use the device to observe laminar flow. The flow will be analyzed with a smartphone camera. For the fabrication, the PDMS monomer is blended with a curing agent and the pre-polymer liquid is poured over the master mold. We will use capillaries as the mold in this experiment. This method is fast, reproducible, and only requires materials, which are commercially available and comparatively inexpensive.

Fluid streams in a microfluidic channel exhibit laminar flow and do not display turbulent mixing. Mixing only occurs toward the outlet, where the two fluid streams combine. Although laminar flow results in separated fluid streams, some diffusion occurs at the interface between the streams. The distance (L_D) a molecule can travel by diffusion is given by:

$$L_D = \sqrt{2Dt} \tag{4.1}$$

where D is the diffusion coefficient of a molecule and t is the time required for diffusion.

This experiment has been adapted from references 2–3 and can be implemented for students from high school to undergraduate levels. The PDMS-based microfluidic device can be designed, created, and tested within a couple of hours, which is compatible with high school and undergraduate chemistry lab schedules. It requires a ~ 5-hour single session or 2.5-hour two sessions. After completing this experiment, students will be able to understand key concepts underlying microfabrication such as device design, fabrication, and testing.

4.2 Design of the Microfluidic Device

As capillary tubes are used as mold, the thickness of the capillary determines the size of the channel. Most common gas chromatography capillary tubes are in the range of $\sim 400 \, \mu m$ OD. Two kinds of channel designs are suggested: T- and acute angle designs as shown in Figure 4.1.

Laboratory Methods in Microfluidics. DOI: http://dx.doi.org/10.1016/B978-0-12-813235-7.00004-0

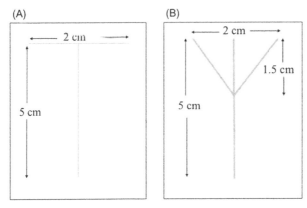

FIGURE 4.1 Design of microfluidic channel. (A) T-design and (B) acute angle design.

4.3 Chemicals and Supplies

PDMS base and curing agent (e.g., Sylgard 184 Silicone Elastomer Kit), red and blue food colors.

Capillary tube, capillary cutting stone, Petri dish, paraffin, tweezers, double-sided tape, plastic cup, wood stick, hot plate, vacuum desiccator, razor blade, biopsy punch (~2-mm diameter), scotch tape, glass microscope slides (75 × 50 mm and 25 × 75 mm), pliers, binder clips, syringe pump, and a phone camera/smartphone.

4.4 Hazards

Lab coats, safety glasses, and gloves should be worn at all times during the course of this experiment. Uncured PDMS mixture should be poured in a well-ventilated area. Measuring the viscous PDMS liquid can be messy. Cover the work surface and the balance with aluminum foil. Extra caution should be taken when using razor blades and biopsy punches. The food colors will stain both skin and cloth.

4.5 Experimental Procedure

A group of 2−3 students is recommended to perform the experiment.

4.5.1 Fabrication

Follow the below steps to fabricate the PDMS microfluidics device:

1. Cut some pieces of capillary tubes using a ceramic cutting stone. After cutting the capillaries, dip their tip at each side in melted paraffin and let it solidify. This helps to prevent air trapped in the capillaries from exiting while degassing the PDMS, which can lead to bubbles and displacement of the capillaries.

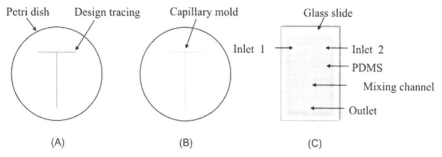

FIGURE 4.2 (A) Petri dish base with channel design tracing, (B) capillary network as mold, (C) a PDMS-glass microchip showing mixing channel, inlets and outlets.

2. Grind the tip of the capillaries, which will be making the connections between capillaries, to remove the bumps using a polishing stone, such as the side surface of a cutting stone, with horizontal movements, to flatten the tip completely.

3. Mark the design of the channel design on a Petri dish (Figure 4.2A). Attach the capillaries on the marking with the help of double-sided tape (Figure 4.2B). Examine the gaps and connections between the capillaries to ensure good contact.

4. Carefully place a small piece of paraffin using sharp tweezers on each of the capillary connections. Heat the tip of the tweezers on a hot plate for a minute and then carefully approach the tip to melt the paraffin, which fills the gaps between the capillaries due to capillary effects and joins them to one another. The excess melted paraffin can be removed carefully with the tip of a razor or sharp tweezers.

5. In a disposable plastic cup, add ~25 g of PDMS elastomer base and the curing agent (known as cross-linker) in a 10:1 (w/w) ratio. Thoroughly mix the mixture for 10 minutes using a wooden stick.

6. Place the mixture in a vacuum desiccator and degas the mixture under reduced pressure for 30 minutes at room temperature. The PDMS will froth while degassing occurs and will then stabilize as a clear liquid on completion of the degassing process.

7. Pour the PDMS prepolymer mixture onto the Petri dish containing capillary mold. Cure the PDMS by placing the casting assembly on a hot plate at 65°C for 2 hours or overnight to polymerize the polymer.

8. Remove the whole piece of cured PDMS from the Petri dish by gently peeling back the edges of the PDMS using razors before trying to remove the entire device from the templates. Make sure the capillaries are also removed from the PDMS piece. The capillaries can be carefully pulled out from the PDMS using pliers. Wear gloves while handling the PDMS piece so that the oils on your hands are not transferred to the PDMS microfluidic device. If residues of paraffin are left inside the microchannel, these can be dissolved and washed by flushing the microchannel with acetone.

9. Cut the PDMS device to the desired shape and size (e.g., 4 cm × 6 cm) using a razor blade. Use this piece to obtain the flat surface on top and place it inside another Petri dish.

10. Using the metal tip of a core punch or biopsy punch, create holes (2 mm diameter) through the PDMS for the inlets and outlet.
11. Use Scotch tape (by gently pressing and peeling) to remove dirt or dust particles from the PDMS mold prior to assembling the device.
12. On a clean 5 × 7 cm glass slide, put down a strip of double-sided tape large enough to seal the footprint of the microfluidic network (Figure 4.2C). Ensure that the tape lies flat against the slide by rolling a clean, dust-free syringe barrel (or some other sturdy cylinder) over the tape. Make sure there are no bubbles between the glass slide and the tape.
13. Form the final device by placing the PDMS piece with channel side facing down onto the double-sided tape on the glass slide. Press gently and evenly to remove any air pockets between the double-sided tape and the PDMS mold. Take care not to collapse the microfluidic channels by pressing too hard. Use binder clips to hold the PDMS and glass slide together.

4.5.2 Visualization of Laminar Flow

Mixing in a microfluidic channel can be visualized by using two different colored solutions of water as follows:

1. Introduce the red color solution into inlet 1 and the blue color solution into inlet 2. If you have a syringe pump (that's better), then introduce the colors at 10 mL/h. You can vary the flow rate to understand its effect on diffusion. A single syringe pump is used to drive both solutions to ensure equal flow rates of each solution in the channel. In the absence of the syringe pump, drop a drop of color into each inlet and do not let it dry by adding color continuously.
2. Take a photo of the chip at different intervals of time (e.g., every 10 seconds) using your smartphone. A microscope attached to a computer is a better option.
3. Clean your device to reuse it by running water through the device.
4. Using ImageJ software, analyze the flow profile across the channel at different locations starting from inlets all the way to outlets. In ImageJ, use three different color channels (the RGB) and see the difference. See if there is mixing toward the outlet. (*See* Appendix V for procedure for using ImageJ.)

4.5.3 Lab Report

Write a lab report following ACS style for journal articles. It should include an abstract, introduction, experimental, results and discussion, and conclusion sections with diagrams, tables, calculations, and images wherever necessary.

Consider the following points when writing the report:

1. Include a short description on the use of PDMS in microfluidics and laminar flow.
2. Outline the major steps and workflow involved in the experiment.
3. Add representative photos of the fluid flow in the channel.

4. Include the gray value profile of channels, especially toward the outlets.
5. Add the effect of flow rate on diffusion.

4.6 Additional Notes[3]

1. Students should be encouraged to vary mold patterns, device dimensions (e.g., chip thickness, channel width, etc.), PDMS curing temperatures (hence, chip rigidity), and substrate materials (glass cover slide or another PDMS slab).
2. An optimal PDMS chip thickness is ~ 2 mm. Thinner chips offer better flexibility to enhance bonding but the thin wall often fails to provide enough support for the needles, which results in leaking at inlets or outlets.
3. The ratio of PDMS and curing agent can be changed depending on the texture of the chip desired. The higher the PDMS %, the harder the chip.
4. If no vacuum desiccator is available to degas PDMS, the sample can be left to be degassed and cured at room temperature overnight followed by postcuring in an oven at 65°C. To increase the speed of PDMS curing, much higher temperatures can be used to cure within a few minutes.
5. It is preferable to use square capillaries because there are no gaps formed at the connection sites due to the square shapes of the capillaries as opposed to the round capillaries, which will have a small gap formed at the connection sites due to the rounded shape of the capillary walls. These gaps can, however, get filled with melted paraffin.
6. While pulling out the capillaries, to make this process easier, the network can be immersed into or washed with acetone, which will swell the PDMS and expand the channels prior to pulling out the capillaries.
7. The PDMS piece can be cleaned by rinsing it in a stream of DI water and ethanol when necessary.

Video Demonstrations:

Mixing PDMS and curing agent: https://www.youtube.com/watch?v=4W067pc_UUE
Degasing PDMS: https://www.youtube.com/watch?v=hDzpnbLRaGQ

4.7 Assessment Questions

1. What is PDMS? Give the structure and list three characteristics of this material.
2. What are some advantages of using PDMS in microfluidics against glass substrate?
3. Briefly describe the molding and device fabrication process used in this experiment.
4. Did two colors mix while flowing through the microfluidic channel? Justify your answer.
5. For a small molecule with a diffusion coefficient of $2 \cdot 10^{-5}$ cm^2/s to diffuse in a channel for 30 seconds, what is the distance this molecule can travel across?

References

1. Anderson JR, Chiu DT, Wu H, Schueller OJ, Whitesides GM. Fabrication of microfluidic systems in poly (dimethylsiloxane). *Electrophoresis*. 2000;21(1):27–40.

2. Hemling M, Crooks JA, Oliver PM, Brenner K, Gilbertson J, Lisensky GC, et al. Microfluidics for high school chemistry students. *J Chem Educ*. 2013;91(1):112–115.

3. Ghorbanian S, Qasaimeh MA, Juncker D. Rapid prototyping of branched microfluidics in PDMS using capillaries. *Chips Tips (Lab Chip)*. 2010; http://blogs.rsc.org/chipsandtips/2010/05/03/rapid-prototyping-of-branched-microfluidics-in-pdms-using-capillaries.

Determination of Electroosmotic Flow in a Glass Microfluidic Device Using a Neutral Marker

5.1 Background

Electroosmotic flow (EOF) is the flow of fluid in a microchannel when electric field is applied across the terminals of the microchannel. In many microfluidic experiments this flow is used to move analytes in the channel. EOF produces a uniform plug profile that results in lesser band broadening compared to pressure-driven flows.[1]

There are various methods[2] to determine the EOF in microfluidic device. In this experiment, we will follow a simple procedure to estimate the EOF velocity in a microchannel of a glass microfluidic device using a neutral fluorescent dye. The EOF velocity will be measured by allowing a neutral dye solution to flow from the injection point to the detection point in the channel by applying an electric field. We will also study the influence of ionic strength and applied electric field on EOF.

An electrically neutral dye has zero electrophoretic velocity. Therefore it migrates at the EOF velocity provided the system has no pressure gradient. Important criteria[2] for a neutral marker to be used in this experiment are:

- Dye should be soluble in background buffer solution.
- Adsorption of the dye to the capillary wall should be negligible.
- The charge on the marker may depend on the pH of the solution. Thus the user should verify that the marker is applicable to the buffer.

The EOF velocity (v_{EOF}) in cm/s is simply calculated from the migration time of a neutral marker for a given distance as given by:

$$v_{EOF} = \frac{l}{t} \tag{5.1}$$

where l is the distance traveled by the flow (distance from injection point to detection point) and t is the time taken for the flow to reach the detector.

The EOF velocity depends on applied electric field (E) and electroosmotic mobility (μ_{EOF}) as given in Eq. (1.1). According to the equation, the velocity is proportional to the applied electric field in V/cm and electroosmotic mobility.

Laboratory Methods in Microfluidics. DOI: http://dx.doi.org/10.1016/B978-0-12-813235-7.00005-2

We can also calculate the electroosmotic mobility in this experiment using:

$$\mu_{EOF} = \frac{L \times l}{V \times t} \tag{5.2}$$

where L is the total length of the channel and V is the applied voltage. Remember that the electric field is the applied voltage per unit distance.

Electroosmosis also depends on the pH and ionic strength of the solution by changing the zeta potential and double layer of the glass surface. Electroosmosis decreases with increasing ionic strength of the solution. In this experiment, we will have to reduce the pressure-driven flow, which is the flow generated by different liquid heads at the reservoirs. This can be done to some extent by using wider reservoirs so that the liquid height in the reservoirs is reduced.

This lab experiment is suitable for courses in microfluidics and engineering, etc., at both undergraduate and graduate levels. It will take one 3-hour lab session if the microchip is provided to the students. After completing this experiment, students will learn to estimate EOF velocity, electrokinetic injection, voltage control, etc.

5.2 Microfluidic Device Design

The glass microchip ($\sim 1'' \times 2''$) is comprised of a T-channel. As an example, the channels can be $\sim 100\,\mu m$ wide and $10\,\mu m$ deep. The main channel can be ~ 3 cm. The detection point should be at ~ 2 cm downstream from the injection area (Fig. 5.1). Glass microchip devices can be fabricated (*see* Chapter 2: Fabrication of a Glass Microfluidic Device) in the

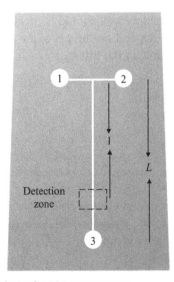

FIGURE 5.1 Design of the microfluidic device for EOF measurement.

lab if the facility is available. If not the microchip can also be purchased from a commercial vendor (*see* Appendix VI).

5.3 Chemicals and Supplies

NaOH, rhodamine B, sodium tetraborate, methanol, and DI water.

Glass microfluidic device, voltage control device, and epifluorescence microscope with camera.

5.4 Hazards

Eye protection, hand gloves, and laboratory coats are recommended. Contaminated materials should be disposed of appropriately as hazardous chemicals. The edges of the glass microfluidic devices pose a small cutting hazard. Special care has to be taken while handling high voltage.

5.5 Experimental Procedure

5.5.1 Solution Preparation

Prepare 100 µM rhodamine B solution separately in the following solvents:

1. DI water
2. 0.1 mM sodium borate
3. 1.0 mM sodium borate
4. 10.0 mM sodium borate

5.5.2 Microfluidic Device Preparation

Make sure the chip is clean and no dust particles are inside the reservoirs and channel. The reservoirs should have larger diameters for easier handling and reducing pressure-driven flow.

Introduce a solution of 1 M NaOH into the entire microchannel network for 30 minutes. Wash the channel and reservoir with water followed by methanol for 10 minutes each.

5.5.3 Data Collection

1. Place the chip on the microscope stage. Turn on the microscope and power supply. Place 200 µL of dye solution in reservoir 1 and fill reservoirs 2 and 3 with the dye-free solvent. The liquid levels in both reservoirs must be equated at this step.
2. Open the LABVIEW program to set up the voltage program for the power supply and injection.

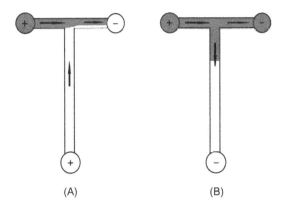

FIGURE 5.2 Schematic of the (A) flow profile set up and (B) injection of the dye.

3. Apply a high voltage to the sample reservoir (200 V) and reservoir 3 (300 V). Ground reservoir 2 for 2 minutes to have the flow profile as shown in Fig. 5.2A.
4. Now inject the dye solution into the long channel by applying 1000 V at reservoir 1 and grounding the two remaining two reservoirs.
5. Right after hitting the injection, start recording the signal at the detection point.
6. Repeat step 4 with 500 V and 1500 V injection voltages.
7. Repeat steps 2–4 using remaining three solutions of the neutral dye.
8. For each electropherogram recorded, find out the time of the dye solution front.
9. Calculate the EOF velocity and mobility.
10. Compare the EOF velocity and mobility at different conditions of buffer and voltage and plot graphs.

5.5.4 Lab Report

Write a lab report following ACS style for journal articles. It should include an abstract, introduction, experimental, results and calculations, discussion, and conclusion sections with diagrams, tables, and images wherever necessary.

Consider the following points when writing the report:

1. Include a short description of electroosmosis on microchip.
2. Outline the major steps and workflow involved in the experiment including the settings for the fluorescence microscope, photomultiplier tube detector, and LABVIEW program used in the measurements.
3. Add a description of the instrumentation involved with schematics.
4. Include representative electropherograms.
5. Estimate the EOF with sample calculation.
6. Determine the reproducibility of dye front migration times.
7. Discuss the effect of voltage and ionic strength on the velocity.

5.6 Additional Notes

1. It is recommended that once students are ready to start the experiment, they should let the TA or instructor inspect it.
2. Instead of a glass device, devices made of other substrates can also be used.
3. The relative voltages given in data-collection sections may not exactly work for your chip. Therefore slight modification of the applied voltage may be required.

References

1. Landers JP, ed. *Handbook of Capillary and Microchip Electrophoresis and Associated Microtechniques.* 3rd ed. New York: CRC Press; 2008.
2. Devasenathipathy S, Santiago JG. *Electrokinetic Flow Diagnostics. Berlin, Heidelberg.* Springer; 2005.

Electrophoretic Separation in a Microchannel

6.1 Background

Electrophoresis is a technique in which charged species under the influence of applied electric field are separated. Among several types of electrophoresis, capillary electrophoresis is widely practiced. In capillary electrophoresis, electric field is applied across the two terminals of submillimeter diameter capillaries. As microfluidic separation systems are now commercially available, electrophoresis in microchip, known as microchip electrophoresis, has become an important tool in modern analytical procedures. Microchannel works as a separation capillary in microchip electrophoresis analysis. It is considered a low-cost, high-throughput, and high-speed method. Microchip electrophoretic separations are carried out with low volumes (nL–pL) of analyte and samples. Microfabricated electrophoresis systems have been used to analyze samples ranging from small molecules, to DNA, to proteins in a variety of biomedical and molecular biology applications.[1]

This laboratory experiment[2,3] involves the separation of a mixture of dye-tagged amino acids. This experiment is suitable for courses that teach analytical techniques such as instrumental analysis and can also be part of molecular biology courses. After the completion of this experiment, students will learn the use and fundamentals of microfluidics, separations, and fluorescence microscopy detection. Students will also be able to identify different peaks in the sample. Furthermore, they will use electropherograms to calculate and explain relevant parameters such as the separation efficiency (theoretical plate number) and reproducibility of the results. Students will also study the correlation of theoretical plates with injection time and separation distance. Details of electrophoretic separation are given in Chapter 1: Introduction to Microfluidics of this book.

This experiment requires a microchip, a solid-state laser, a high-voltage power supply, and a fluorescence microscope. Fluorescent light from sample molecules is collected by photomultiplier tube (PMT) detection. The PMT signal is converted to voltage, filtered with a preamplifier, and digitized with an analog-to-digital converter to enable collection as detector signal vs time using a designed LabView program.

A fluorescence detector at the end of the separation channel of the microfluidic device records the fluorescence of analyte bands to obtain an electropherogram. For laboratories lacking access to lasers, a UV source coupled with a filter cube can also be used for fluorescence detection.

A well-defined plug of sample mixture has to be introduced into separation channel. Injection is one of the key elements that determines the quality of separation of analytes. There are a number of injection strategies in microfluidic electrophoresis such as pinched

Laboratory Methods in Microfluidics. DOI: http://dx.doi.org/10.1016/B978-0-12-813235-7.00006-4

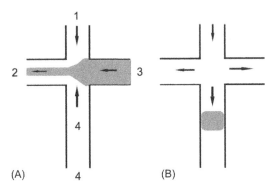

FIGURE 6.1 Schematic of pinched injection scheme. The pointed arrows indicate fluid flow direction. The darker color indicates sample solution. (A) Sample loading and (B) sample injection.

injection and gated injection. This lab will follow a *"pinched"* injection scheme (Fig. 6.1). In this injection scheme, the nL−pL volume of the sample is injected into the separation channel. This is the most commonly used electrokinetic injection method and allows for well-defined, fixed-volume, and reproducible injection of a sample plug.[4]

As shown in Fig. 6.1, sample solution is kept in reservoir 3 and the rest of the reservoirs are filled with buffer solution. Then, the waste reservoir (i.e., 2) is grounded and the remaining reservoirs are applied with high voltage in such a way that the flows from these three reservoirs are directed toward the waste reservoir. This step is known as the flow profile set-up step. In a second intermediate step, high voltage at reservoirs 1 and 4 are relaxed for a second or so (injection time) to allow the intersection of channel to be filled with the sample. Finally, in the injection step, reservoirs 2, 3, and 4 are grounded while keeping high voltage at reservoir 1 until the analyte reach the detection point.

Amino acids are labeled with a fluorophore and the detection of separated analyte bands is performed with a fluorescence detector directly on the microchip. A mixture of three amino acids (Fig. 6.2) consisting of a positive, a negative, and a neutral species will be separated in this experiment. These amino acids are glycine, aspartic acid, and arginine. Glycine is a neutral species in aqueous solution, aspartic acid is anionic, and arginine is cationic.

This lab experiment is designed for two periods each consisting of roughly 3 hours of experimentation. In the first lab session, the instructor demonstrates the experiment and students prepare dye-tagged amino-acid sample. In the second lab session, students carry out their independent experiments. It is recommended that a group of three students perform this experiment.

6.2 Design of the Microfluidic Device

The design of microchip electrophoresis systems (Fig. 6.3) consists of a sample injection zone, an electrophoresis separation channel, and a system for detection of the migrating analytes. The microchip device is about $2'' \times 1''$ in size in which the channels can be

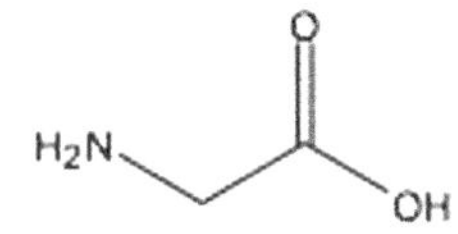

Glycine

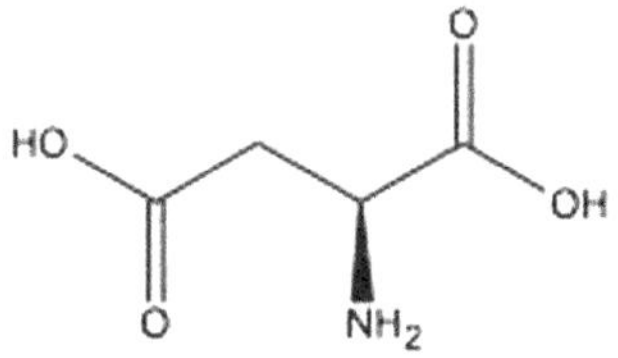

Aspartic acid

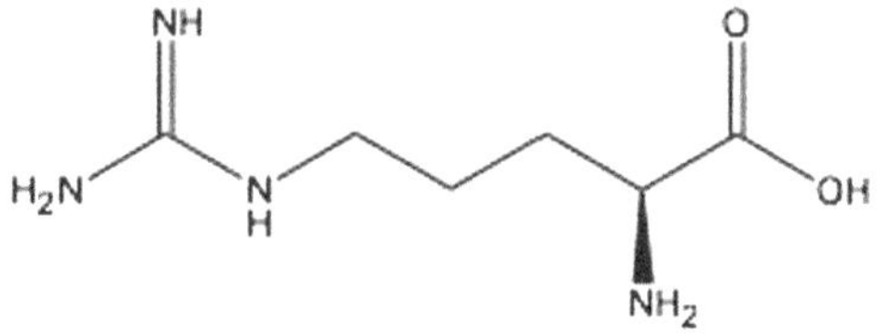

Arginine

FIGURE 6.2 Structure of the three amino acids used in the experiment.

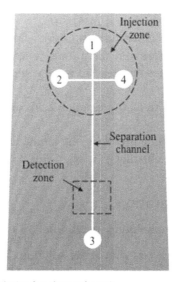

FIGURE 6.3 Design of the microfluidic device for electrophoresis.

~100 μm wide and 10 μm deep. A commercial microchip (*see* Appendix VI) can be purchased for this experiment or a poly(methyl methacrylate) (PMMA) or glass device can be fabricated in the lab if facility is available.

6.3 Chemicals and Supplies

DI water, fluorescein-5-isothiocyanate (FITC), amino acids (e.g., glycine, aspartic acid, and arginine), hydroxypropyl cellulose. The fluorescent compounds are light sensitive, so store them in the dark when you are not using them.

Microchip, power supply system, and fluorescence microscope system.

6.4 Hazards

Microchip electrophoresis requires high voltage. Be careful about the laser radiation. Be advised of the involved risks of accidental shocks. Lab coats, gloves, and eye protection should be worn throughout the experiments. Contaminated materials and chips should be disposed of appropriately as hazardous chemicals.

6.5 Experimental Procedure

6.5.1 Sample Preparation

The amino acids are labeled with FITC as follows:

1. Prepare buffer solution (borate buffer—1 mM).
2. Combine 200 μL of 6 mM FITC in dimethyl sulfoxide with 600 μL of a 3 mM solution of the amino acid. Leave the mixture to react at room temperature in the dark for more than 24 hours. Even longer times may lead to more complete reaction and elimination of unconjugated FITC peak.
3. Repeat the above procedure with all three amino acids and a mixture of three amino acids.
4. Use hydroxypropyl cellulose while preparing the amino acid solution to reduce the electroosmotic flow inside the channel.

6.5.2 Electrophoretic Separation

Follow the below steps for electrophoretic separation of dye tagged amino acid sample.

1. Fill all the channels and reservoirs with buffer solution. Make sure there are no air bubbles inside the channels and the buffer solution levels in the reservoirs are the same.
2. Turn on the microscope and detection system.

Reservoir	Voltage applied (V)		
	Flow profile	Injection	Separation
1	Intermediate	Intermediate	High
2	High	High	Grounded
3	Intermediate	Grounded	Grounded
4	Grounded	Grounded	Grounded

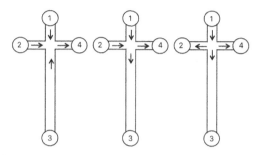

FIGURE 6.4 Voltage programming for electrophoretic separation (top) and schematic of the fluid flow at different steps of the experiment (bottom).

3. Place the device on the microscope stage. Secure the device in position with 2−3 pieces of tape. Verify that the high-voltage power supplies are off, and then carefully insert the four platinum electrodes into their respective reservoirs.

4. Focus the microscope at the intersection of the microchannels.

5. Replace the buffer in reservoir 2 with sample solution (choose one amino acid).

6. Check that the electrodes are still in position, and verify that the inject switch is in the "inject" position. Then, turn on the high-voltage power supplies. After 1−2 seconds, the fluorescent sample will flow through the injection channel, and the PMT will detect the fluorescence signal. Adjust the fine focus, and then turn off the high voltage and the power supplies. (Students must verify that all wires are in the appropriate reservoirs before turning on the power supplies, and must always ensure that the power supplies are off before touching any wires) (Fig. 6.4).

7. Move the microscope lens to a detection spot that is ∼2 cm down the separation channel.

8. Then, carry out an injection/separation. Be sure to hit the "record" button at the same time you switch the box from "inject" to "run."

9. The voltage will drive your sample electrophoretically through the injection channel, and you should see a gradually increasing signal at the detector.

10. Obtain three replicate separations for each sample.

11. Repeat steps 5−10 with remaining two amino acids and a mixture of amino acids.

12. Perform the separation of amino acids by varying the injection time and detection position. Try at least two more injection times and detection points (one lower and higher than the ones you used before).

13. At the end of each lab period, switch off the high voltage and turn off the power supplies and microscope. Flush your chip with water, vacuum the channels dry and store it with the channels dry.

6.5.3 Switching Samples in the Microchip

1. Shut off the power supplies. Switch the side knob to the "eye" position. Carefully remove the electrodes from the reservoirs and pull off the tape securing your device.
2. If your device is working fine, you can just pipet out old buffer from each reservoir and replace it with new buffer. Then, pipet out your old sample from the reservoir, rinse the reservoir with buffer, and pipet the new sample into the reservoir. Your device is now ready for additional separations.
3. If your device is having problems, it is best to flush it out and refill it with buffer. Spray each reservoir with a wash bottle and then pipet the liquid out. Repeat as needed. Apply vacuum to each reservoir to remove the liquid from the channels.

6.5.4 Data Analysis

1. Open individual amino acid electropherogram and estimate the migration time and peak width.
2. Open the mixed amino acid sample and estimate the migration time and peak width for each peak.
3. Compare the migration time from the mixed sample with individual ones and identify the amino acids in the mixture.

6.5.5 Lab Report

Write a lab report following ACS style for journal articles. It should include an abstract, introduction, experimental, results and calculations, discussion, and conclusion sections with diagrams, tables, and images wherever necessary.

Consider the following points when writing the report:

1. Include a short description of electrophoretic separation in microchip including its strengths and weakness and the difference to conventional CE.
2. Outline the major steps and workflow involved in the experiment including the settings for the fluorescence microscope, PMT detector, and LABVIEW program used in the measurements.
3. Add a description of the instrumentation involved with schematics.
4. Include representative electropherograms of amino acids with labels.
5. Estimate flow velocity, electrophoretic mobility of each amino acids, plate height, and separation resolution.
6. Determine the reproducibility of peak migration times and theoretical plate counts.

7. Add the effect of injection time and separation distance on the number of theoretical plates.

8. Show the representative calculations for theoretical plate count.

6.6 Additional Notes

The instructor can choose the ratio of amino acids in the mixture to be separated. The mixture should contain enough individual amino acid to be detected by the detector.

6.7 Assessment Questions

1. What are the differences between microchip electrophoresis and capillary electrophoresis?

2. How did the injection time affect the separation?

3. How did the detection point affect the separation?

References

1. Landers JP, ed. *Handbook of Capillary and Microchip Electrophoresis and Associated Microtechniques.* 3rd ed. New York: CRC Press, Taylor & Francis Group; 2008.

2. A microchip capillary electrophoresis experiment for the instrumental analysis laboratory [Internet]; 2014.

3. Chao TC, Bhattacharya S, Ros A. Microfluidic gel electrophoresis in the undergraduate laboratory applied to food analysis. *J Chem Educ.* 2011;89(1):125−129.

4. García CD, Chumbimuni-Torres KY, Carrilho E. *Capillary Electrophoresis and Microchip Capillary Electrophoresis: Principles, Applications, and Limitations.* New Jersey: John Wiley & Sons; 2013. Available from: http://dx.doi.org/10.1002/9781118530009.

A Simple Experiment for the Study of Droplet Microfluidics

7.1 Background

Droplet-based microfluidics[1] is a subset of microfluidics in which discrete volumes of dispersed phase known as microdroplets are created by combining two or more immiscible fluids. The most commonly used two fluids are aqueous (water) and organic (mineral oil). Either of the two fluids can be used as dispersed phase (droplets) or continuous phase (carrier). The size of droplets typically ranges from 1 μm to 100s of μm (femto- to nanoliter range). The use of a microfluidic device allows generating highly controlled and reproducible discrete monodisperse droplets. Each droplet can be independently controlled, transported, mixed, and analyzed. Thus the produced droplets are utilized for a diverse range of chemical and biological assays[2] such as:

- Encapsulation of cells, DNA, or magnetic beads for research, analysis, and diagnostics.
- Protein crystallization, polymerase chain reaction-based analysis, enzyme kinetics.
- Drug delivery via polymer particles and drug formulation.
- Bulk precision manufacturing of emulsions and foams for foods and cosmetics.
- Nanoparticles, paints, and polymer particles synthesis.
- Microreactors for performing chemical and biochemical reactions.

Among the various methods for generating microdroplets in a microfluidic device, T-junction and flow-focusing are two popular methods. In this experiment, we will use the T-junction method in which the inlet channel containing the dispersed phase perpendicularly intersects the main channel that contains the continuous phase (*see* Fig. 7.1). The two phases form an interface at the junction, and as fluid flow continues, the tip of the dispersed phase enters the main channel. The shear forces generated by the continuous phase and the subsequent pressure gradient cause the head of the dispersed phase to elongate into the main channel until the neck of the dispersed phase thins and eventually breaks the stream into a droplet. The sizes of the droplets depend on the fluid flow rates, channel widths, and viscosity of the two phases.[1]

In this experiment, we will generate and characterize the microfluidic droplets in terms of generation rate and size. We will use a T-junction microfluidic device and an oil-in-water emulsion. The syringe pump will be used to control the fluid flow in channel and control the droplets size. The microfluidic droplet generator will then be mounted on a bench top and video will be recorded using a smartphone. Thus recorded video will be analyzed using image-processing software, ImageJ, for estimating generation rate (droplets per second, or Hz). The diameter of droplets (μm) will be back calculated from known area measurements.

Laboratory Methods in Microfluidics. DOI: http://dx.doi.org/10.1016/B978-0-12-813235-7.00007-6

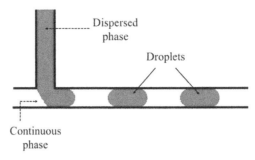

FIGURE 7.1 Schematic of a microfluidic system for generating droplets.

A μm per pixel conversion factor can be obtained from the known width of wooden coffee stirrer that has been used for device fabrication. The mean diameter of plugs formed in the device can then be estimated by using this factor and height of channel in pixels. Monodisprse droplets are those that have low standard deviation relative to the mean diameter.

This lab experiment may be implemented in graduate and undergraduate courses in physics, chemistry, and engineering. In addition, it can also be demonstrated in high-school courses. Students can work in groups of 2−4 depending on class size. This can be completed in a 2-hour lab session provided the microfluidic chip is ready. More time would be required if the lab involves the fabrication of the device.

7.2 Microfluidic Device Design

In this experiment, we will use a T-junction PDMS microfluidic device (Fig. 7.1).

7.3 Chemicals and Supplies

PDMS, Sylgard 184 silicone elastomer, blunt 18G needle, glass slides, camera phone such as Samsung or iPhone, ImageJ, deionized water, blue food dye, syringe pumps, coffee stirrers (600 μm in width or less), and light mineral oil.

7.4 Hazards

Lab coats, gloves, and eye protection should be worn as a precaution during device fabrication.

7.5 Experimental Procedure

7.5.1 Solution Preparation

Prepare a blue food color solution in appropriate amount.

7.5.2 Fabrication of Device (Optional)[3]

Fabrication of T-junction PDMS chip by using wooden coffee stirrers as master mold.

- Cut coffee stirrers (600 μm in width) to unequal lengths using household scissors or a blade.
- Glue flat to the base of a Petri dish at a 90° angle in order to form the junction. Allow drying.
- Mix the PDMS and curing agent in a 10:1 ratio.
- Pour the mixture over the mold and leave to cure in a 70°C temperature-controlled dry oven.
- Peel off the solidified chip off the mold and bond to a glass slide.
- Punch inlets and outlet using a blunt 18G needle.
- Connect the inlets to separate syringes in a syringe pump.

7.5.3 Droplet Generation

1. Fill up the syringes with dispersed and continuous phase solutions.
2. Apply pressure using the syringe pump and start the flow.
3. Record the flow using a smartphone.
4. Vary the pressure of dispersed and continuous phases.
5. Repeat steps 2−3 with different pressure.

7.5.4 Image Analysis and Quantitation

1. Using the recorded video, count the number of droplets per unit time.
2. By tracking the leading edge of the plug traveling in the channel over time, determine the approximate velocity of the flow.
3. Take a screenshot of the video and then measure the dimension of the droplet.
4. By using a known dimension stirrer and ImageJ estimate the micrometer per pixel conversion factor. Then estimate the size of the droplet.

7.5.5 Lab Report

Write a lab report following ACS style for journal articles. It should include an abstract, introduction, experimental, results and discussion, and conclusion sections with diagrams, tables, calculations, and images wherever necessary.

Consider the following points when writing the report:

1. Include a short description of the principles behind droplet microfluidics.
2. Outline the major steps and workflow involved in the experiment. Be sure to report the experimental conditions for the assay and the operational parameters for the instrumentations used.

3. Add representative photos of the channel showing droplets.
4. Include results with representative calculations of flow rate, droplet size, droplet generation rate, etc.

7.6 Additional Notes

1. If syringe pumps are not available, hand pumps can be used.
2. You can use other disperse phase and continuous phase solutions.

7.7 Assessment Questions

1. What is a droplet? How does a plug differ?
2. How do flow-focusing and T-junction methods different?
3. How do you confirm a laminar flow in this experiment? (Estimate the Reynolds number. A low Reynolds number means laminar flow.)

References

1. Teh SY, Lin R, Hung LH, Lee AP. Droplet microfluidics. *Lab Chip*. 2008;8(2):198−220.

2. Chou WL, Lee PY, Yang CL, Huang WY, Lin YS. Recent advances in applications of droplet microfluidics. *Micromachines*. 2015;6(9):1249−1271.

3. Bardin D, Lee AP. Low-cost experimentation for the study of droplet microfluidics. *Lab Chip*. 2014;14 (20):3978−3986.

8

Laminar Flow and Diffusion in a Microchannel

8.1 Background

When two or more fluids are allowed to flow in microchannel side by side, mixing of fluids results almost entirely by diffusion. This allows dissimilar fluids to flow alongside each other in a microchannel over long distances without significant mixing. Such flow is known as laminar flow and is characterized by low Reynolds number (R_e),[1] which is a dimensionless parameter without a physical unit. This number relates inertial to viscous forces in fluids and is given by:

$$R_e = \rho U_o L_o / \eta \tag{8.1}$$

where ρ is the fluid density (kg/m^3), U_o is the characteristic velocity, L_o is the typical length scale, and η is the shear viscosity. A high R_e number indicates a turbulent flow with inertial forces dominating. In contrast, low R_e number indicates laminar flow with viscous forces dominating. This number is used to differentiate the type of flow in the microfluidic device. Typically, the transition from laminar to turbulent flow in round pipes occurs at $R_e = 2000-3000$.

In this lab, we will follow a very simple and low-cost method to observe and understand laminar flow and diffusion in microfluidic device. A Y-channel Jell-O chip[2] will be used to realize the laminar flow and the flow will be made possible by gravity without using expensive syringe pumps. Jell-O is gelatinous material and has been used to fabricate microfluidic devices for various educational and demonstration purposes. The flow in such setting can be directly visualized without the use of a microscope and can be recorded using a smartphone camera.

Another parameter we are going to cover in this experiment is diffusion. Diffusion is a fundamental transport phenomenon that is random in nature. The transport occurs from high concentration to low concentration. Diffusion of a molecule is characterized by a term called molecular diffusion coefficient (D), which depends on the size of molecule, viscosity of solution, temperature, etc. The distance (X_D) traveled by molecules because of diffusion can be calculated as follows, where t represents time of travel:

$$X_D = \sqrt{2Dt} \tag{8.2}$$

Another dimensionless parameter, known as the Péclet number (Pe), is also useful in understanding diffusion, and is found by:

$$Pe = U_o L_o / D \tag{8.3}$$

Laboratory Methods in Microfluidics. DOI: http://dx.doi.org/10.1016/B978-0-12-813235-7.00008-8

where U_o is the characteristic velocity, L_o is the typical length scale, and D is the diffusion coefficient, which is given by Eq. (8.4) assuming the particle is spherical:

$$D = k_B T / 6\pi\eta\alpha \qquad (8.4)$$

where k_B is the Boltzmann constant, T is the temperature, η is the shear viscosity, and α is the molecule size. The diffusion coefficient of typical food coloring dye is about 200 $\mu m^2/s$. Larger Pe indicates convective mass transfer domination and little diffusion along the length of the channel. Smaller Pe indicates diffusion. Therefore using Re, we can calculate if the devices have laminar or turbulent flow and using Pe, we can determine whether convective mass transfer or diffusion dominates.[2]

The diffusivity of a dye in water is greater than that of the same dye in oil. We will test diffusion of dye molecules in water and oil. In addition, we will test with different concentrations of dye solutions. Finally we will test dye solution in water and in glucose water. The diffusion of dye molecules across the channel can be observed and measured by determining the transverse concentration profile across the laminar flow at multiple locations along the channel. In laminar flow, the concentration profile is proportional to the square root of the distance from the junction. The concentration gradient is very sharp immediately after the two liquids meet and then smooths out further down the channel.

The flow in this experiment is initiated and controlled by syringe pumps, known as pressure-driven flow, which is different from the flow generated by applying electric field as described in Chapter 5: Determination of Electroosmotic Flow in a Glass Microfluidic Device Using a Neutral Marker.

This experiment[2] can be completed in a 2-hour lab session. Students will fabricate a microfluidic device from easily available material like Jell-O, perform pressure-driven flow in microchannel, estimate flow velocity, and estimate diffusion of dye molecules.

8.2 Microfluidic Device Design

The microfluidic chip in this experiment is Y-shaped. The channels in the Jell-O chips are defined by the size of the wooden stirrers. The outlet channel is about 3″ long and two inlet channels are 1″ long each.

8.3 Chemicals and Supplies

Deionized (DI) water, blue food color, oil, sugar, flavored Jell-O jelly powder, and original gelatin.

Styrofoam plates, wooden coffee stirrers, scotch tape, double-sided tape, aluminum plate/pan, drinking straw, no-stick cooking spray, smartphone, 10 mL syringes, disposable transfer pipets, and stopwatch.

8.4 Hazards

Personal protective equipment including lab coat, goggles, and gloves are recommended.

8.5 Experimental Procedure

8.5.1 Solution Preparation

The following solutions and reagents are required for this experiment:

1. At least three different concentrations of dye in water.
2. Dye solution in water containing glucose.
3. Jell-O and gelatin mixture: Dissolve two pouches of Jell-O jelly powder (~ 8 g) in 120 mL of purified water in a beaker and dissolve one pouch of gelatin powder in the same amount of water in a second beaker. Then place the beaker containing the Jell-O solution on a hot plate and heat it to a boil. Remove this beaker from the heat and add the gelatin solution to it. Heat the mixture again to a boil and remove from the heat.

8.5.2 Fabrication of Jell-O Chip

1. Cut the coffee stirrers into desirable shapes and sizes.
2. Adhere single-side tape onto stirrer surface to produce a smoother surface.
3. Attach the stirrers onto a foam plate using double-sided tape to create a Y-pattern as shown in Fig. 8.1.
4. Spray no-stick oil spray on the foam plate. This will help peel off the chip later.
5. Pour the liquid of Jell-O and gelatin mixture onto the molds.
6. Leave it to cure for 2 days in a refrigerator.
7. Remove the chip from the refrigerator. Peel the chip from the mold and place it in an aluminum dish. The high sugar content from the Jell-O and gelatin mixture provides a natural seal on the aluminum dishes used here.
8. Punch two inlet and one outlet hole using a drinking straw with a gentle twisting motion.

8.5.3 Laminar Flow Setup

1. Attach two 10-mL syringes in the inlets.
2. Load the first syringe with blue water and the second one with clear water.

FIGURE 8.1 The Y-channel design of the Jell-O chip (left). The photo of the chip on the right shows laminar flow in the channel by flowing blue-color dye solution from the inlet on the right and flowing water from the inlet on the left. The Reynolds number for this chip was found to be 30. *(The photo on the right was reprinted with permission from Yang CWT, Ouellet E, Lagally ET. Using inexpensive Jell-O chips for hands-on microfluidics education.* Anal Chem. *2010;82(13):5408–5414. Copyright (2010) American Chemical Society.)*

3. Inject both solutions simultaneously.
4. Visualize the flow with your eyes.
5. Record the flow using your smartphone. Also take close-up photos of the channel.
6. Repeat steps 2–5 by varying the fluid height in the inlets to vary the flow rate.
7. Repeat steps 2–5 by using oil and dye water.
8. Repeat steps 2–5 by using different concentrations of dye solution and water.

8.5.4 Lab Report

Write a lab report following ACS style for journal articles. It should include an abstract, introduction, experimental, results and discussion, and conclusion sections with diagrams, tables, calculations, and images wherever necessary.

Consider the following points when writing the report:

1. Include a brief summary of the relevant theory providing a background to your experiment.
2. Outline the major steps and workflow involved in the experiment.
3. Add a description of the accessories needed.
4. Include representative images and videos.
5. Estimate the Re with sample calculation.
6. Show representative calculations for Pe.

8.6 Additional Notes

1. While gravity feed is simple to implement, it is also very sensitive to external disturbances, such as table vibrations. Be aware that even atmospheric disturbances (caused by breathing) can affect your results. Gravity feed can also be difficult to start.
2. Curing the chips overnight is usually sufficient. However, curing for at least two days in the refrigerator results in more robust Jell-O chips.
3. To prevent leakage from the inlets, the syringes should be perfectly perpendicular with the surface, and the head of the syringes should completely seal the inlets.
4. Both clear and green-colored water should be injected slowly but evenly into the channels in order to create the laminar flow profile.
5. The most common problem encountered with the Y-Channel chip is with peeling the chip off of the mold, especially at the junction of the three coffee stirrers. Patience, care, and experience will help with the peeling process.
6. A simple T-Channel chip instead of an Y-channel chip can also be used.

8.7 Assessment Questions

1. Estimate the Re for a chip with a 0.3 cm-wide channel. The density of water is 1.0 g/cm^3 and the viscosity is 0.010 g/cm. Assume that the fluid flow velocity is 1.0 cm/s.

2. What happens to the flow rate when you change the size of the channel?

3. What are the differences between pressure-driven and electrokinetics-driven flow profile in a microchannel?

4. Calculate the flow rate (μL/min) that results in a mean-flow velocity of 1 mm/s in the long channel.

5. What is the maximum flow velocity in the channel then?

6. Consider dye molecules with a diffusion coefficient of 2.5×10^{-5} cm^2/s. How much time will it take to travel from one side to other of a 200 μm-wide microchannel?

References

1. Kirby BJ. *Micro- and Nanoscale Fluid Mechanics: Transport in Microfluidic Devices.* New York: Cambridge University Press; 2010.

2. Yang CWT, Ouellet E, Lagally ET. Using inexpensive Jell-O chips for hands-on microfluidics education. *Anal Chem.* 2010;82(13):5408−5414.

9

Beer's Law Using a Smartphone and Paper Device

9.1 Background

When light is passed through a sample, molecules in the sample absorb a certain amount of light. The rest of the light passes through the sample. Amount of light absorbed by the sample is explained by Beer−Lambert law, commonly known as Beer's law. This law is commonly used in high school and undergraduate chemistry lab courses for the determination of concentration of an absorbing analyte.

Absorbance (A) is expressed in logarithmic terms as given by:

$$A = - \log(I/I_0) \tag{9.1}$$

where I_0 is the initial intensity of light and I is the intensity of same light after it passes through the sample (*see* Fig. 9.1). The absorbance is also expressed as Eq. (9.2) and is known popularly as Beer's law or Beer−Lambert's law:

$$A = \varepsilon \, l \, c \tag{9.2}$$

The molar absorption coefficient (ε) is a constant and is a wavelength-dependent intrinsic property of the chemical species. It is a measurement of how strongly a chemical species absorbs light at a given wavelength with units of L/mol/cm. The l and c in Eq. 9.2 refer to the light-path length (cm) and analyze concentration (mol/L), respectively.

According to Eq. (9.2), the amount of light absorbed by a sample is proportional to the concentration of the absorbing species. The more concentrated a sample is, the greater the amount of light absorbed. When we plot the measured absorbance of various concentrations of the chemical species against corresponding concentrations, we get a calibration curve. The best-fit equation of the calibration curve is then used to determine the concentrations of unknown sample.

In this experiment, we will verify Beer's law using a smartphone or digital camera or any mobile phone camera combined with image processing software such as ImageJ. As the use of smartphones by students is widespread, introducing such experiments into undergraduate/high-school laboratory experiments generates a great deal of enthusiasm and discussion. In this experiment, we will measure the color intensity on a *μPAD*.

The color of light absorbed by a substance and the color we see are different. A substance that absorbs all wavelengths in the visible range appears black to our eyes. In contrast, a substance absorbing none of the incident visible light appears white (all light reflected) or colorless (all light transmitted) to our eyes. For example, the indigo dye in blue jeans has a

Laboratory Methods in Microfluidics. DOI: http://dx.doi.org/10.1016/B978-0-12-813235-7.00009-X

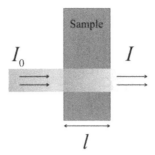

FIGURE 9.1 Schematic showing light passing through a sample (light path $= l$). I_0 is initial intensity of light before passing through the sample and I is the intensity of same light after it passed through the sample.

Table 9.1   Colors Absorbed and Corresponding Colors Observed by our Eyes

Wavelength Absorbed (nm)	Color Absorbed	Color Observed
410	Violet	Yellow-green
430	Blue-violet	Yellow
480	Blue	Orange
500	Blue-green	Red
530	Green	Purple
560	Yellow-green	Violet
580	Yellow	Blue-violet
610	Orange	Blue
680	Red	Blue-green
720	Red-purple	Green

maximum absorbance in the 500−650 nm range. This absorbance is in the red-to-green region. The not-absorbed wavelength lies in 400−500 nm range and gives rise to blue-violet color. An aqueous solution that appears yellow, with a narrow range around 550 nm, means that wavelengths on either side of yellow, primarily blues, greens, and reds, are being absorbed. A green solution would be expected to transmit green wavelengths, while blocking blues, yellows, and reds. Table 9.1[1] provides a summary of the relationships between the wavelengths of colors absorbed and the colors observed.

In the RGB (red, green, blue) color model, any color is considered to be a combination of red, green, and blue light. Considering this fact, we will use three different color solutions: nickel nitrate, cobalt nitrate, and starch−iodine complex. We will make calibration curves for all three colors to verify Beer's law. In addition, we will further use the regression equation for starch−iodine complex to estimate the amount of starch present in potato.

After completing this experiment, students will be able to make serial dilution of solutions, analyze images using ImageJ software, make calibration curves, estimate limits of detection, and determine the amount of starch in potato. This method can be applied to a wide range of laboratory settings, both in research and in education. It is suitable for general

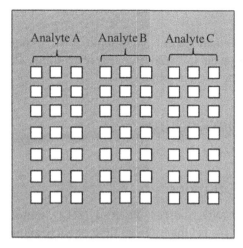

FIGURE 9.2 Design of paper microfluidic device used in this experiment. Triplicate measurements are taken for each analyte.

chemistry to organic chemistry to water and environmental chemistry courses. It can be completed in a 3-hour laboratory session.

9.2 Microfluidic Device Design

This experiment uses a paper microfluidic device ($\sim$10 cm $\times$ 10 cm) with a very simple design. Each device consists of seven rows and nine columns of test zones (5 $\times$ 5 mm) (*see* Fig. 9.2). Three columns are used for triplicate measurement of each analyte. The five rows in each column are used for five standard solutions and the sixth one is used for blank. The seventh row should be used for the unknown potato starch sample. This device can be created using a variety of methods: wax-printing method (*see* Chapter 10: Acid–Base Titrations on Paper and Chapter 11: Simultaneous Determination of Protein and Glucose in Urine Sample Using Paper-Based Bioanalytical Device), permanent-marker method (*see* Chapter 13: Determination of Nitrite Ions in Water Using Paper Analytical Device), and wax-pen method (*see* Chapter 12: Quantitative Determination of Total Amino acids in Tea Using Paper Microfluidics and a Smartphone). An appropriate method can be chosen based on the availability of resources. The device can be fabricated by the student or the instructor can provide the device.

9.3 Chemicals and Supplies

Deionized (DI) water, $Co(NO_3)_2 \cdot 6H_2O$, $Ni(NO_3)_2 \cdot 6H_2O$, cornstarch, tincture iodine, and potato.

Paper device, micropipettes and tips, electric juicer, conical-bottom screw-cap plastic centrifuge tubes, smartphone/digital camera, and ImageJ software.

9.4 Hazards

Tincture of iodine contains ethanol and is flammable. It is also poisonous and should not be ingested. Care should be taken to avoid staining of clothing. Protective eye cover, safety glasses, and gloves should be worn at all times.

9.5 Experimental Procedure

Work in pairs on this experiment.

9.5.1 Preparation of Solutions

Obtain stock solutions of $Co(NO_3)_2 \cdot 6H_2O$, $Ni(NO_3)_2 \cdot 6H_2O$, standard cornstarch, and potato samples from instructor or TA. Estimate the volume of each solution required for the assays and prepare the following solutions:

1. *Standard solutions of Co^{++} and Ni^{++} salts*: (12.5, 10.0, 7.50, 5.00, and 2.50 μM) by diluting the 25.0 μM stock solution with water. Use DI water as blank solution.
2. *Starch stock solution*: Prepare starch stock solution using cornstarch. At first prepare starch slurry by mixing 1 g cornstarch with some water. Then slowly pour the slurry and mix into 75 mL of water that had just been boiled. Bring the volume to 100 mL by further addition of water to form a stock solution of 10 g/L. Prepare successive dilutions of this starch stock solution using water.
3. *Starch standard working solutions*: Perform serial dilution of the stock-starch solution to get at least five different concentrations ranging from 10 g/L to 0.0005 g/L.
4. *Iodine indicator*: Add 50 μL of tincture of iodine per 3 mL of water (approximately one part tincture of iodine per 60 parts water).

9.5.2 Isolation of Starch from Potato[2]

1. Remove the skin of a potato and weigh it.
2. Use an electric "juicer" to produce potato juice. Collect the potato juice in two 50-mL conical-bottom screw-cap plastic centrifuge tubes. Leave the tubes for 2 hours. The solids content and liquid are readily separated upon standing. There are two layers of solids: a dense white solid (starch) and a brown layer (likely to be cell wall components) of lesser density. The dense white solid of starch becomes tightly packed upon standing.
3. Separate the fuliginous liquid portion of the juice from the solid components by decanting.
4. Remove the brown solids from the mixture of solids using two washing cycles: resuspend the combined solids in cold tap water by vigorous shaking of the tube and let the mixture settle two more hours. Carefully resuspend the less dense brown solids in the liquid fraction by slow rotation. Invert the tube to remove the brown solid and liquid suspension; the white solid is obtained at the bottom of the tube.

5. Stir the white solid with a spoon to make it wet paste. Take 1 g of wet paste to form 100 mL of a potato starch. Successive dilutions in water of this solution may be required.

9.5.3 Color Assay

1. Pipet 5 μL of cobalt nitrate standard solutions and drop into the assay regions in the first column of Analyte A in the paper device. Repeat this procedure for columns two and three. Drop just water into the sixth-row assay regions.
2. Repeat step 1 for nickel nitrate and cornstarch standards. The last row under Analyte C should be used for the potato starch sample.
3. Add 10 μL of the iodine solution in all assay regions under Analyte C.
4. Let the assay regions dry. It may take ~10 minutes.

9.5.4 Image Acquisition and Analysis

1. Take a photo of the entire microfluidic device using a smartphone/phone camera. Make sure the camera of the phone is well focused.
2. Measure the gray intensity in each square assay region in the microfluidic device using ImageJ software (*see* Appendix V).
3. Split the image into red, green, and blue channel. Measure the gray intensities again in each channel.

9.5.5 Data Analysis

1. Collect the gray intensity values into Microsoft Excel.
2. Make Beer's law plots of intensity vs concentration using the collected data for all channels and the raw image for each color solution.
3. Include the best-fit line and equation for your data.
4. Compare the correlation coefficient and slope among different channels for each color. Identify the best channel for each color solution.
5. Use the regression equation from the cornstarch calibration curve and intensity value of the potato starch sample to estimate the concentration of starch in the potato.
6. Estimate the error in your reported value.

9.5.6 Lab Report

Write a lab report following ACS style for journal articles. It should include an abstract, introduction, experimental, results and discussion, and conclusion sections with diagrams, tables, calculations, and images wherever necessary.

Consider the following points when writing the report:

1. Include a short description of Beer's law using the paper device and smartphone including its strengths and weakness compared to a conventional spectrophotometric method.

2. Outline the major steps and workflow involved in the experiment.
3. Add representative photos of the paper device and calibration curves.
4. Estimate the concentrations of the unknown potato starch sample and show representative calculations.

9.6 Additional Notes

1. Capillary pipettes can be used if micropipettes are unavailable.
2. Care should be taken to minimize reflection and shadow on the device while taking an image.
3. Using a background color that is complementary to the observed solution color should produce the most quantitative results.
4. It is not necessary to only use the chemicals (colors) given in this experiment. Any food color dye can be used to demonstrate the same idea.

9.7 Assessment Questions

1. Do your solutions obey Beer's law? How do you know?
2. Provide the equation for the best-fit regression line obtained for your Beer's law plots for all analytes and compare the sensitivity (i.e., slope of the regression line).
3. What are the limitations of Beer's law?
4. What are the advantages and disadvantages of this method compared to the conventional spectrophotometric method?
5. What is the concentration of an unknown solution whose absorbance value was measured to be 0.46: The regression equation for this solution is as follows:

$Y = 4.0890\,X + 0.004\,4.0890.$

References

1. Harris DC. *Quantitative Chemical Analysis*. 8th ed. New York: W.H. Freeman and Company; 2010.
2. Mathews KR, Landmark JD, Stickle DF. Quantitative assay for starch by colorimetry using a desktop scanner. *J Chem Educ.* 2004;81(5):702.

Acid−Base Titrations on Paper

10.1 Background

Titrations such as acid−base, redox, etc., are widely practiced in teaching and analytical laboratories to determine the concentration of an unknown. This method, unlike modern instrumental analysis, does not require calibration curve. For example, in acid−base titration, the concentration of an acid or a base solution is directly calculated from the volume required to reach the end point. In addition, titration permits on-site measurement using no instrumentation since the human eye determines the end point. The currently used titration methods involve a large volume of reagents and samples, glassware (burette and pipettes), technical skills, and a long period of time.[1]

In this experiment, which is adapted from reference 2, we will fabricate microfluidic paper-based analytical devices (μPADs) using a wax-printing method and carry out acid−base titration. This μPAD titration method has more advantages than classic titrimetry in terms of cost, speed, portability, and disposability. The proposed μPAD is a viable alternative to classic titration methods in the analysis of several chemical species at field sites where glassware and large volumes of solutions are difficult to obtain. Since μPADs are inexpensive, this new titration technique may replace traditional titrations, especially in developing countries.[3]

The wax-printed μPADs used in the experiment consist of central sample loading zone, reaction reservoirs, and detection reservoirs. The reaction reservoirs are modified with various amounts of a primary standard solution of potassium hydrogen phthalate (KHPth), whereas the detection reservoirs are modified with a constant amount of indicator. When a small amount of base such as sodium hydroxide (NaOH) solution as a sample is applied, it wicks to the reaction reservoir and reacts with the acid (Eq. (10.1)). If there is extra base in the sample, it moves to the detection reservoir and by reacting with the indicator, it gives a characteristic color. The number of detection reservoirs with no color change can be used to determine the concentration of base in the sample. In this case, the concentration of analyte is determined directly with no calibration curve. In this experiment, students will carry out two different titration experiments: one involving the determination of base using primary standard acid (NaOH vs KHPth) and another involving the determination of acid using primary standard base as in the case of the determination for HCl using Na_2CO_3 as primary standard substance (Eq. (10.2)).

$$KHC_8H_4O_4 + NaOH \longrightarrow KNaC_8H_4O_4 + H_2O \qquad (10.1)$$

$$2HCl + Na_2CO_3 \longrightarrow 2NaCl + H_2O + CO_2 \qquad (10.2)$$

This experiment is suitable for general and analytical chemistry courses and can be completed in a 3-hour laboratory session. After completing this experiment, students will be able

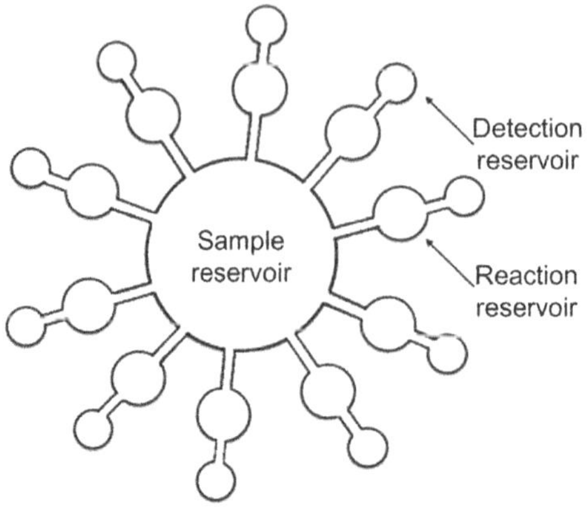

FIGURE 10.1 Design of the paper device for acid–base titration.

to fabricate a paper device using wax printing, perform acid–base titration, and perform image processing using software like ImageJ.

10.2 Design of the Microfluidic Device

The titration μPAD (30 × 30 mm) consists of a sample reservoir located at the center (10 mm) and ten reaction (3 mm) and detection reservoirs (2 mm) each arranged radially (Fig. 10.1).

10.3 Chemicals and Supplies

Deionized (DI) water, NaOH, KHPth, HCl, Na_2CO_3, and phenolphthalein.

Whatman #1 filter paper, wax printer, hot plate, clear packaging tape, micropipette tips, design software, and ImageJ software.

10.4 Hazards

Always wear safety goggles and gloves in the laboratory, especially when working with strong acids and the bases.

10.5 Experimental Procedure

10.5.1 Designing and Fabricating Titration Paper Device Using Wax Printing Method

Following fabrication steps (Fig. 10.2) are recommended to be carried out in group.

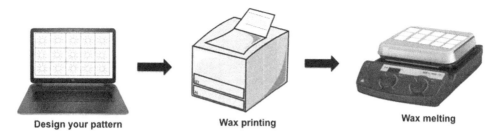

FIGURE 10.2 Schematic of wax printing method procedure for fabricating paper-device. Design of the paper device for acid–base titration.

1. Design the *μPAD* as shown in Fig. 10.1 using Microsoft Office PowerPoint or other design software available in your lab. Print the design on a sheet of Whatman filter paper using a wax printer.
2. Heat the wax printed paper at 150°C for 2 minutes on a hot plate.
3. Cover the backside of the printing surface with clear packing tape to prevent solution from leaking out underneath the *μPAD*.
4. Measure the areas of the channels and reservoirs using ImageJ.
5. Cut each *μPAD* in 30 × 30 mm individual pieces.

10.5.2 Solution Preparation

Obtain an unknown sample of NaOH and HCl from your instructor or TA. Prepare the following solutions in DI water except when stated otherwise. Estimate the volume of these solutions required for this experiment before preparing the solutions.

1. Stock solution of 1 M Na_2CO_3.
2. Stock solution of 1 M KHPth.
3. Make nine more dilutions of the above primary solutions ranging from 0.01 M to 0.1 M.
4. Stock solution of 1% (w/w) phenolphthalein in ethanol and a 0.5% dilution working solution.

10.5.3 Titration

1. Apply 1 μL of different concentrations of primary standard solutions (KHPth for NaOH determination) to reaction reservoirs using micropipette.
2. Apply 1 μL of indicator solution (0.5% phenolphthalein) to detection reservoirs using micropipette.
3. Introduce 30 μL of sample solution (NaOH) onto the sample loading zone using a micropipette. Let the sample solution wick to the reaction zone and then detection zones. This takes about 2 minutes. Observe the color changes in the detection reservoirs (*see* Fig. 10.3 as example). Take a photo of the device after the sample solution occupies whole detection zones.

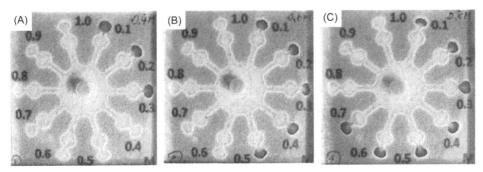

FIGURE 10.3 Examples of titration (NaOH vs KHPth using 0.15% phenolphthalein). Numbers on the sides of the detection zones indicate concentration of primary standard. (A) 0.4 M NaOH, (B) 0.6 M NaOH, and (C) 0.8 M NaOH. *(Reprinted with permission from Karita S, Kaneta T. Acid—base titrations using microfluidic paper-based analytical devices. Anal Chem. 2014;86(24):12108—12114. Copyright (2014) American Chemical Society.)*

4. Repeat this procedure in a second device to determine the concentration of unknown HCl by using Na_2CO_3 as the primary standard.
5. Use your eyes and ImageJ software (*see* Appendix V) to find out which detection zone showed a color change.
6. Estimate the concentration of unknown based on the volume of unknown, concentration of primary standard solution, and its volume used. Using ImageJ software estimate the gray value of each detection zone and explain the result you see.

10.5.4 Lab Report

Write a lab report following ACS style for journal articles. It should include an abstract, introduction, experimental, results and discussion, and conclusion sections with diagrams, tables, calculations, and images wherever necessary.

Consider the following points when writing the report:

1. Include a short description of acid—base reactions in the paper device including its strengths and weaknesses.
2. Outline the major steps in the experiment.
3. Add representative photos of the titration paper device.
4. Estimate the concentration of unknowns using both visualization by eye and ImageJ software. Show representative calculations.

10.6 Additional Notes

1. Generally students fabricate the device and use it for acid—base titration in the same lab session. If they have to store the device after applying standard and indicator, it is better to store at room temperature under light. For more than a week storage, it is better to store in refrigerator.

2. The volumes needed to fill the reaction and detection reservoirs may vary depending on type of filter paper, temperature, and humidity. In that case, it is necessary to optimize the applied volumes of the reagents first using dye solution. It should be noted that the amount of phenolphthalein is an important factor in clear visualization of a color change. Therefore concentration and volume of indicator (phenolphthalein) may also have to be optimized.
3. Same strategy can be used to determine nitric acid, sulfuric acid, acetic acid, and ammonia, etc. with appropriate standards and indicators.
4. NaOH and HCl solutions ranging from 0.01 M to 0.1 M can be given to students as unknown.

10.7 Assessment Questions

1. What is the molar concentration of H_3O^+ in a solution whose pH is 2.1?
2. Suppose 1 μL of 0.05 M KHPth is needed to see the color change. How many moles of NaOH are present in your 30 μL unknown sample? Show the calculation.
3. What indicator is used in this experiment? Write its structure and describe the color change of the indicator.
4. Calculate the concentration of 40 μL base used required to neutralize 2 μL of 0.9 M acid.

References

1. Harris DC. *Quantitative Chemical Analysis*. 8th ed. New York: W.H. Freeman and Company; 2010.
2. Karita S, Kaneta T. Acid–base titrations using microfluidic paper-based analytical devices. *Anal Chem.* 2014;86(24):12108–12114.
3. Myers NM, Kernisan EN, Lieberman M. Lab on paper: iodometric titration on a printed card. *Anal Chem.* 2015;87(7):3764–3770.

Simultaneous Determination of Protein and Glucose in Urine Sample Using a Paper-Based Bioanalytical Device

11.1 Background

Urinalysis[1] is one of the most widely used diagnostic methods to diagnose a number of diseases in clinical laboratories. It is especially useful as a screening test for the detection of renal and other diseases.

Glucose and protein are two target analytes among many other chemicals analyzed in urine. Normally these substances do not appear in urine in detectable amounts. So their appearance in urine shows a pathological condition. For example, protein appears in the urine during renal disease. Proteins play a central role in cell function and cell structure. Since protein molecules are very large, they are not normally present in measurable amounts in the urine. The detection of protein in urine, called proteinuria,[3] may indicate renal infections or kidney diseases. Glucose appears in the urine during diabetes mellitus. Glucose, a monosaccharide, is the principal sugar in blood, serving the tissues as a major metabolic fuel and does not normally appear in urine. When sugar appears in the urine, it points to diabetes mellitus. Hence urine sugar tests are extremely useful in the detection of the disease and monitoring the effectiveness of the diabetic treatment.[2] The presence of a detectable amount of glucose in the urine (15−20 mg/dL) is known as glycosuria.

11.1.1 Protein Assay

Some pH indicators change color in the presence of protein. The protein assay in this experiment[4] is based on the nonspecific binding of tetrabromophenol blue (TBPB) to proteins. TBPB (Fig. 11.1) binds to proteins through a combination of electrostatic (sulfonate) and hydrophobic (biaryl quinone methide) interactions. When bound, the phenol in TBPB deprotonates, and the color of the dye shifts from yellow to blue. TBPB is an acid−base indicator in the pH range 3.0−4.6. The acidic form appears yellow and the basic form appears blue. TBPB also appears blue when bound to protein. The development of any green to blue color indicates the presence of protein. The intensity of the color is proportional to the amount of protein present. The protein-complex color will begin to fade, so be sure to get a photo and make observations in a timely manner.

Laboratory Methods in Microfluidics. DOI: http://dx.doi.org/10.1016/B978-0-12-813235-7.00011-8

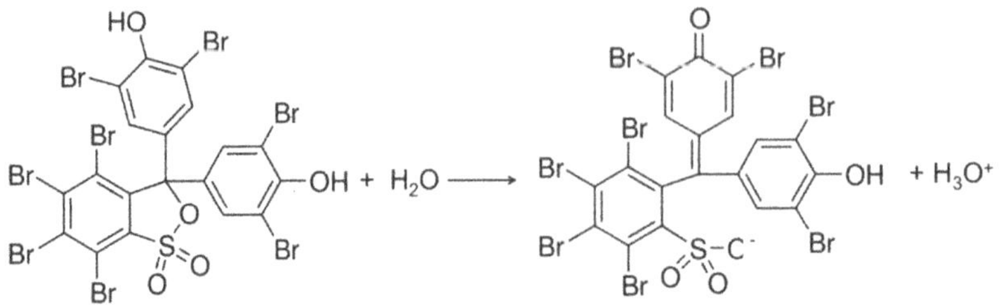

FIGURE 11.1 Acid–base equilibrium of tetrabromophenol blue. The acidic form appears yellow and the basic form appears blue. TBPB also appears blue when bound to protein.

11.1.2 Glucose Assay

The colorimetric glucose assay used in this experiment is based on a double sequential enzyme reaction.[5] Glucose oxidase (GOx) enzyme catalyzes the formation of gluconic acid and hydrogen peroxide from the oxidation of glucose. A second enzyme, horseradish peroxidase (HRP), is then used to catalyze the reaction of hydrogen peroxide with a potassium iodide chromogen to give iodine. This assay is used in many dipstick-based enzymatic tests specific for glucose. The associated change in color from colorless to brown (I^- to I_2) indicates the presence of glucose. The glucose zones in the paper device will turn brown/orange as triiodide (I^{3-}) is formed in a redox reaction.

$$\text{Glucose} \underset{\text{GOx}}{\rightarrow} H_2O_2 + \text{gluconic acid} \tag{11.1}$$

$$H_2O_2 + 3I^- + 2I^- \underset{\text{HRP}}{\rightarrow} I^{3-}(\text{brown}) + 2H_2O \tag{11.2}$$

In this experiment,[5] students will simultaneously perform two clinically important colorimetric bioassays, one for glucose and one for protein, in an artificial urine sample using a single paper-based microfluidic device (μPAD) and a smartphone. The amount of glucose and protein in the sample is calculated by comparing the signal of the sample to that of the standards. The $\mu PADs$ can be prepared using a wax-printing method. Other fabrication methods can be used too. After patterning the paper, reagents for assays are added to the microfluidic device. Once the reagents are dry, the $\mu PADs$ are ready for use. The sample is then added to the device and is allowed to completely react with the reagents. A photo of the test device is taken and color change that occurred from mixing for quantitation is measured. Visual inspection of color in the assay region provides qualitative information. Using a camera as a "detector," a quantitative value can be determined.

This experiment is suitable for undergraduate and in some cases high-school students of chemistry, biochemistry, clinical biochemistry, medical laboratory technology, etc. Students can work alone or in pairs depending on the class size. The entire experiment can be completed in one 3-hour laboratory session. After completing this experiment, students will be able to learn techniques in clinical chemistry and bioanalysis, along with the emerging field of microfluidics.

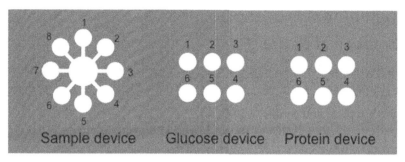

FIGURE 11.2 Design of a paper-based multiplex microfluidic device.

11.2 Microfluidic Device Design

A strip of three *µPAD* contains (Fig. 11.2): individual assay sets for sample, glucose standards, and protein standards. The sample device consists of a central sample loading zone connected to eight detection zones by microchannels. The glucose and protein standard devices contain six independent test zones. Each device is approximately 3 cm in diameter. The sample loading zone in the sample device is $\sim$5 mm in diameter. The remaining detection zones in the sample device and test zones in the glucose and protein devices are $\sim$3 mm in diameter. An extra set of devices can be printed for backup. The sample will be introduced in the center and the hydrophobic channels will guide the solution through the channel to the detection zones on the periphery. The sample device allows for triplicate measurement of both glucose and protein along with blanks for both analytes.

11.3 Chemicals and Supplies

DI water, glucose reagent solution, protein reagent solution, artificial urine solution, glucose stock standard, protein stock standard, priming solution, citrate buffer with 2% Tween 20, and TBPB.

Hot plate, aluminum foil, micropipettes and tips (2.5, 100, and 1000 µL), vortex, Whatman #1 chromatography paper, wax printer, centrifuge tubes (1.5 mL and 5 mL), a timer, and ImageJ software.

11.4 Hazards

Protective eyewear must be worn at all times. TBPB is a dye that will stain skin and clothing. The used paper devices can be thrown in the trash. Turn in any unused devices to your instructor. Return excess unknown solution, glucose reagent, priming solution, and TBPB to your instructor for disposal.

11.5 Experimental Procedure

11.5.1 Fabrication of Device

The wax-printing method for paper-device fabrication is outlined below. However, other methods can also be used.

1. Turn on the hot plate and set it to 150°C.
2. Design your microfluidic device as shown in Fig. 11.2 using Microsoft PowerPoint or other drawing software.
3. Print the design on a sheet of Whatman filter paper.
4. Place your sheet of paper on the oven rack. Allow the sheet to remain in the oven for 90 seconds. The heat melts the wax, allowing the wax to spread through the paper to make a hydrophobic barrier that will contain solutions. The heat also spreads the wax laterally so the channels will look smaller than printed. Turn the paper over and check that the ink has spread through as shown in Fig. 11.3.
5. Cut the devices from the sheet individually or in groups.

11.5.2 Solution Preparations

Obtain the following solutions from the instructor or TA:

- 5 mL of artificial urine solution (AU);
- 500 µL of glucose standard solution (25 mM);
- 500 µL of glucose reagent solution (contains 5:1 solution of GOx–HRP, 0.6 M KI, and 0.3 M trehalose in a pH 6.0 phosphate buffer);
- 500 µL of protein standard solution (2.0 mg/mL, bovin serum albumin (BSA)),
- priming solution (92% water, 8% ethanol v/v, and 750 mM citrate buffer (pH 1.8) with 2% w/v Tween 20);
- 100 µL of protein reagent solution (95% ethanol, 5% water v/v, and 9 mM TBPB); and
- 100 µL of unknown sample.
- Blank is the artificial urine solution without glucose and protein.

Most of the solutions need to be stored in a refrigerator or freezer. Before using, allow them to warm up to room temperature and then mix using a vortex mixer. Determine the volume of each standard required to prepare the working standard solutions.

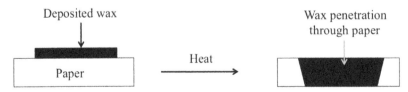

FIGURE 11.3 Penetration of wax through the paper.

Prepare the following solutions:

- *Standard glucose solutions*: 0.25, 1, 2, 4, 6, and 8 mg/dL.
- *Standard protein solutions*: 1.2, 2.5, 5, 10, 15, and 20 mg/dL.

11.5.3 Bioassays

1. Drop 0.50 μL of glucose reagent solution into the detection zones 4, 5, and 6 in both blank and sample devices and in all test zones in glucose device.
2. Similarly, spot 0.50 μL of priming solution into the detection zones 1, 2, and 8 in both blank and sample devices and in all test zones in the protein device. This solution keeps the pH sufficiently low so that the protein reagent, TBPB, remains yellow. Allow these solutions to dry for 5 minutes at room temperature.
3. Next, spot each "primed" protein detection/test zones with 0.30 μL of protein reagent solution (TBPB).
4. Allow the devices to dry for 10 minutes in the dark (in a drawer or under a piece of aluminum foil) at room temperature.
5. Use a micropipette to deliver 14 μL each of the appropriate blank and unknown sample solutions to the center of the blank and sample devices.
6. Deliver a series of glucose standard solutions onto the test zones of the glucose device. Similarly, deliver a series of protein standard solutions onto the test zones of the protein device.
7. Start the timer after putting the samples and standard solutions.
8. Observe the fluid travel through the channels and notice the color changes occurring in the test zones after a few minutes.
9. Looking at the relative colors of the test zones on your devices, estimate the presence or absence of glucose and protein in your unknown sample. Document your observations in your notebook.
10. Obtain a photo of each device using your cell phone.
11. The protein color will begin to fade, so be sure to get a photo and make observations in a timely manner.
12. Analyze the detection/test zones using ImageJ software (*see* Appendix V) for quantitation. Make calibration curves for glucose and proteins.
13. Use regression equations of each of the analytes to estimate their amount in your sample.

11.5.4 Lab Report

Write a lab report following ACS style for journal articles. It should include an abstract, introduction, experimental, results and discussion, and conclusion sections with diagrams, tables, calculations, and images wherever necessary.

Consider the following points when writing your report:

1. Include a short description of the determination of glucose and proteins using colorimetric methods including advantages and disadvantages of using paper devices.

2. Outline the major steps and workflow involved in the experiment.
3. Add representative photos of the device and calibration curve.
4. Estimate the result of concentration of glucose and protein in the unknown urine sample.

11.6 Additional Notes

1. Do not deliver TBPB in the channel. If the priming solution, which acts as a buffer, is not in the channel, a color change will occur even with pure AU solution since TBPB is an acid–base indicator. The buffer lowers the pH and prevents any color change from occurring due to pH. If this happens in one of your trials, do not include that test in your results. If necessary, adjust the volume of TBPB used to spot each zone. Perform the spotting until you have at least three properly spotted circles.
2. To ensure that the solution remains on each paper device and does not go through onto the benchtop, set the device on a test tube rack or weighing boat. The back of the device can be taped with packing tape. You can remove the devices from the weighing boat once they are dry.
3. Suggested unknown concentrations: unknown #1: 0.70 mM glucose and 100 mg/dL protein; unknown #2: 2.50 mM glucose and 20 mg/dL protein.
4. A GC syringe can also be used for the spotting instead of a micropipette.
5. When spotting the test zones, students can adjust the micropipettes to optimize the reagent spotting.
6. Ensure that the TBPB is stored in the dark. Devices spotted with TBPB can be covered in aluminum foil while drying.

11.7 Assessment Questions

1. Which part of your device is hydrophilic? Which part is hydrophobic? Use this information to explain why the aqueous solution is able to flow through the channels on the device.
2. The glucose assay contains the following reagents: GOx, HRP, potassium iodide, and trehalose.
   a. What is the role of each reagent?
   b. The color change results from a redox reaction between hydrogen peroxide and iodide (products are water and the brown-colored tri-iodide I^{3-}). Write the half reactions and the balanced overall reaction.
3. TBPB is not only used as a dye that binds to proteins, but also is an acid–base indicator.
   a. What is the color (yellow or blue) of followings?
      i. Acidic form of TBPB
      ii. Basic form of TBPB
      iii. Protein-bound form of TBPB
      iv. Unbound form of TBPB

References

1. Free AH, Free HM. Urinalysis, critical discipline of clinical science. *CRC Crit Rev Clin Lab Sci.* 1972; 3(4):481−531.

2. Scherstén B, Fritz H. Subnormal levels of glucose in urine: a sign of urinary tract infection. *JAMA.* 1967;201(12):949−952.

3. Blaine J. *Proteinuria: Basic Mechanisms, Pathophysiology and Clinical Relevance.* Cham, Switzerland: Springer; 2016.

4. Flores R. A rapid and reproducible assay for quantitative estimation of proteins using bromophenol blue. *Anal Biochem.* 1978;88(2):605−611.

5. Gross EM, Clevenger ME, Neuville CJ, Parker KA. In: Hou H, ed. *A Bioanalytical Microfluidics Experiment for Undergraduate Students, in Teaching Bioanalytical Chemistry.* Washington DC: American Chemical Society; 2013.

12

Quantitative Determination of Total Amino Acids in Tea Using Paper Microfluidics and a Smartphone

12.1 Background

Amino acids are the building blocks of all proteins. These acids possess an amine group, a carboxylic acid group, and a varying side chain that differs between different amino acids.

One of the commonly used tests for amino acid determination is the ninhydrin test. In the pH range of 4−8, amino acids react with ninhydrin: 2,2-Dihydroxyindane-1,3-dione (triketohydrindene hydrate). The alpha amino groups of the free amino acids react with ninhydrin to form a blue to purple color change (Fig. 12.1). Ninhydrin reagent is a powerful oxidizing agent that causes the oxidative deamination of the alpha amino acids to form aldehydes and reduced aldehyde forms of ninhydrin. In the process, ammonia and carbon dioxide are released. The ammonia formed in this process then reacts with the ninhydrin as well as the reduced products to form a diketohydrin. This diketohydrin forms a blue-purple colored substance. The color intensity produced is proportional to the amino acid concentration. This test[1] is utilized to identify amines in particular alpha amino acids present in the solution. It is commonly used to detect the lysine in fingerprints.

In this experiment, adapted from reference 2, we will determine the total amino acid content in tea leaf using the ninhydrin test. We will carry out the test on microfluidic paper-based analytical devices ($\mu PADs$), which will be fabricated on a piece of filter paper with a wax pen. The wax-pen method of fabrication is easy and inexpensive. Instead of using a spectrophotometer for the quantitation of the test, we will take a photo of the colorimetric assay reaction with a phone camera. The image will then be analyzed using ImageJ or any other image-processing software. The gray value for the concentration of each amino acid is used to obtain a calibration curve. We will use glutamic acid as a standard solution of total amino acid because it is abundant in tea-leaf extract. The concentration of amino acids in the tea-leaf extract is then calculated using the regression equation of the calibration curve. The total amino acid content ($\mu g/g$) in tea leaves is calculated using the concentration of amino acid in tea-leaf extract (C), volume of tea-leaf extract (V), and weighed mass of tea leaves (m).

$$w = \frac{CV}{m} \tag{12.1}$$

Laboratory Methods in Microfluidics. DOI: http://dx.doi.org/10.1016/B978-0-12-813235-7.00012-X

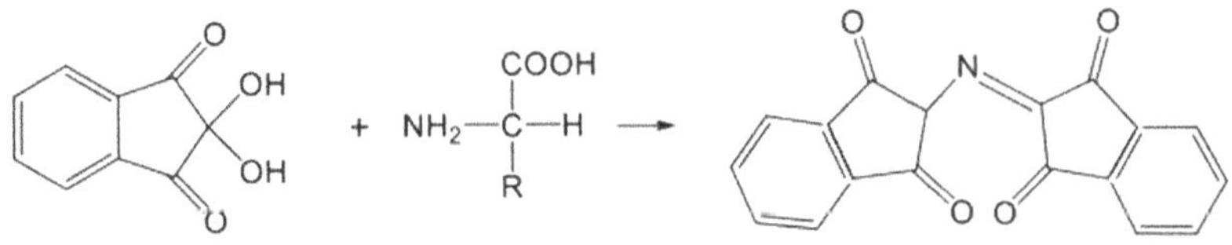

FIGURE 12.1 Reaction involved in the ninhydrin test.

Ninhydrin   Amino acid   Diketohydrin

FIGURE 12.2 Design of the paper device for the ninhydrin test.

This experiment can be completed in 3 hours. This laboratory activity is appropriate for use in an undergraduate course of analytical chemistry, biochemistry, biotechnology, etc. After completing this experiment, students will learn how to fabricate the paper device using a wax pen, carry out an amino acid test using ninhydrin, obtain the calibration curve, and determine the concentration of an unknown sample.

12.2 Microfluidic Device Design

The paper-based microfluidic device is fabricated with a piece of filter paper and a wax pen. The device (6 $\times$ 6 cm) consists of seven detection reservoirs (6 mm) and a central sample-loading zone (8 mm) connected by distribution channels (2 mm wide) (*see* Fig. 12.2).

All distribution channels should have the same length and width and all detection zones the same diameter to ensure the solution spotted in the circle zone flows equally along all six channels and into the detection zones. Thus the detection reservoirs contain equal amounts of ninhydrin after adding the ninhydrin solution in the device. Five detection reservoirs are spotted with various concentrations of glutamic acid as standard amino acid, with the sixth one being the tea-extract sample and the last one the blank solution (no glutamic acid).

12.3 Chemicals and Supplies

Deionized water, ninhydrin, $SnCl_2 \cdot 2H_2O$, glutamic acid, $Na_2HPO_4 \cdot 12H_2O$, $NaH_2PO_4 \cdot 2H_2O$, and tea leaves.

Whatman No. 1 paper, pH meter, hot plate, clear packaging tape, micropipette and tips, wax pen, smartphone, and ImageJ software.

12.4 Hazards

The ninhydrin solution is a strong oxidizing agent. It will react with the proteins of skin. Additionally, ninhydrin is harmful to the eyes, skin, and respiration system. Ingestion of ninhydrin is harmful. Proper caution should be exercised in handling this solution. Students should wear goggles, protective gloves, and long-sleeve lab coats. It is suggested that the reagents and tea extraction should be carried out under fume hood.

12.5 Experimental Procedure

12.5.1 Preparation of Reagents and Solutions

Obtain 1.0 mg/mL stock standard solution of glutamic acid prepared in deionized water. Prepare the following solutions. Estimate the volume of these solutions needed before preparing them.

1. 0.3 M phosphate buffer of pH 8 (*see* Appendix II for procedure).
2. Working glutamic acid standard solutions (0, 10, 30, 60, 90, and 120 µg/mL) by appropriate dilution of the stock solution in the phosphate buffer solution.
3. 2.0% Ninhydrin (*see* Appendix II for procedure).

12.5.2 Tea-Leaf Extraction

1. Weigh 1.0 g of tea leaves accurately into a beaker containing 300 mL of boiling water on a hot plate.
2. Boil the water for 40 minutes to extract the amino acids from the tea leaves.
3. Cool the solution to room temperature by placing the beaker into running cold water.
4. Filter the content and transfer the filtrate into a volumetric flask and dilute to 500 mL with phosphate buffer solution.

12.5.3 Fabrication of Device

1. Hand draw the microchannel pattern in a filter paper using a wax pen.
2. Put the patterned paper in the hot plate (135°C) for 30 seconds. Remove from hot plate and allow it to cool to room temperature.

12.5.4 Determination of Amino Acids

Follow the steps given below to determine amino acids using ninhydrin test (Fig. 12.3).

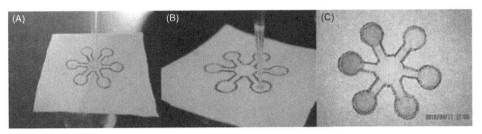

FIGURE 12.3 Procedure for amino acid analysis. (A) Applying ninhydrin reagent, (B) adding standards and samples, and (C) photo of the device after color development. *Reprinted with permission from Cai L, Wu Y, Xu C, Chen Z. A simple paper-based microfluidic device for the determination of the total amino acid content in a tea leaf extract. J Chem Educ. 2012;90(2):232–234. Copyright (2012) American Chemical Society.*

1. Place a beaker of ~ 40 mL capacity on the experimental table, and then put the dry fabricated paper-based device on the top of the beaker.
2. Spot 15 μL of 2.0% ninhydrin on the circle zone in the center of the dry paper-based device. The ninhydrin solution flows into the detection zones within ~ 100 seconds.
3. Allow the device to dry (~ 9 minutes).
4. Spot 1.0 μL of each six standard solutions glutamic acid (0, 10, 30, 60, 90, 120 μg/mL) and the tea sample solution onto the seventh detection zone (one solution for each zone). Let the solutions dry for 5 minutes at room temperature.
5. Heat the device on a hot plate at 80°C for 15 minutes such that the amino acid reacts with the ninhydrin to form a purple-colored complex in the detection zones.
6. Take an image of the device covering all detection zones using a smartphone or digital camera.
7. Analyze the test zones for the mean gray value using ImageJ software (*see* Appendix V).
8. Obtain a calibration curve. The final mean gray values of standards and tea sample used while creating the calibration curve are obtained by subtracting the gray value of blank.
9. Obtain linear equation for the correlation.
10. Calculate the total concentration of amino acids in sample solution according to the mean gray value in the sample detection zone and the linear equation.
11. Estimate the total amino acids content in the given tea leaf in μg/g of leaf.

12.5.5 Lab Report

Write a lab report following ACS style for journal articles. It should include an abstract, introduction, experimental, results and discussion, and conclusion sections with diagrams, tables, calculations, and images wherever necessary.

Consider the following points when writing the report:

1. Include a short description of the determination of amino acids using colorimetric methods including advantages and disadvantages of using paper device.
2. Outline the major steps and workflow involved in the experiment.

3. Add representative photos of the device and calibration curve.

4. Estimate the result of concentration of total amino acids in the tea leaves.

12.6 Additional Notes

1. The instructor may provide a template of device design to students. Students can also use a regular pencil to trace the design pattern onto the filter paper.

2. Wax melting temperature and duration may vary. Therefore it can be changed based on the paper and wax type students use.

3. Even though green tea is suggested as the sample in this experiment, other types of tea such as black and white tea could also be used.

4. It is better to fabricate the paper device when the water is kept boiling for 40 minutes in the sample preparation part, which could save time to ensure that the whole experiment can be completed within 3 hours.

12.7 Assessment Questions

1. Why was the sample cooled to room temperature before measuring the absorbance?

2. What properties of the ninhydrin−amino acid reaction other than color can be used to detect the presence of amino acids?

3. What are the disadvantages of using ninhydrin reagent to estimate total amino acid content?

References

1. Moore S, Stein WH. A modified ninhydrin reagent for the photometric determination of amino acids and related compounds. *J Biol Chem*. 1954;211:907−913.

2. Cai L, Wu Y, Xu C, Chen Z. A simple paper-based microfluidic device for the determination of the total amino acid content in a tea leaf extract. *J Chem Educ*. 2012;90(2):232−234.

13

Determination of Nitrite Ions in Water Using Paper Analytical Device

13.1 Background

Nitrite ions exist naturally throughout the environment but are hazardous to human health.[1] This chemical species is a common pollutant of rivers, streams, lakes, and water supplies. Anthropogenic sources of nitrite include intensive use of chemical nitrogenous fertilizers, improper disposal of plant and animal waste, municipal and industrial wastewater discharge, sewage disposal systems, landfills, etc. Excess nitrite ion can be fatal to infants by causing a condition known as methemoglobinemia. Nitrite also has a significant carcinogenic effect as it can make carcinogenic nitrosamine and *N*-nitroso compounds. Due to concerns about its potential hazards, most international organizations set legal limits for nitrite content in drinking water. For this reason, nitrite levels are routinely determined in quality control analyses of drinking, waste, marine, and underground waters, among others. Sodium nitrite is added to meat products to keep them fresh for an extended period in a cold but not frozen state. This additive gives the meat a pinkish-reddish "cosmetic" color and prevents it from turning brown. In addition, it prevents the growth of *Clostridium botulinum*, the microorganism that produces the deadly botulism toxin. It has been found that feeding sizeable amounts of sodium nitrite to animals results in formation of compounds called nitrosamines, which are suspected of being carcinogenic (i.e., cancer causing).[1] The US EPA limit for NO_2^--N in drinking water is 1 mg/L.[2]

In this 3-hour experiment, which is based on reference 3, we will fabricate a microfluidic paper analytical device (μPAD) and use it for the determination of nitrite ion in both a pre-prepared unknown sample and a pond-water sample. We will follow a simple and convenient fabrication method using a permanent marker pen. A colorimetric assay is employed on the μPAD to determine nitrite ion concentration. A colorimetric reagent for nitrite ion determination is introduced into the microfluidic test zones, followed by standards and sample. After reaction, the color change is easily observed by the naked eye. Finally, a phone camera is used as the transducer to convert the chemical information held in the paper device to quantitative data. Analysis of the photo using image-processing software will generate a calibration curve based on the concentrations of the standard solutions and their corresponding gray values. This calibration curve is then used for the determination of the concentration of an unknown sample.

We will detect and quantify nitrite ions using one of the most common spectrophotometric methods, the Griess test.[4] Griess reagent is an aqueous solution of 0.1% naphthylethylenediamine dihydrochloride and acidic 1% sulfonamide. The nitrite ion reacts with sulfonamide by diazotization, followed by the coupling reaction between the intermediate

Laboratory Methods in Microfluidics. DOI: http://dx.doi.org/10.1016/B978-0-12-813235-7.00013-1
83

and naphthylethylenediamine. This causes a color change of the solution from colorless to red (Fig. 13.1). The diazotization reaction between nitrite ions and the indicator performed on the *μPAD* should be protected from light. The intensity of the color increases as the concentration of added nitrite ion increases.

We will use the paper-based Griess method to determine the nitrite concentration in a simulated water sample as well as in pond water. The regular method will be compared with standard addition method. In the standard addition method, standard solutions of analyte are directly added to the aliquots of the sample. This method is useful when the sample matrix contributes to the analytical signal—the matrix effect. A regression equation from a calibration curve created using a sample spiked with standard solution is used to estimate the concentration of nitrite ions (Fig. 13.2).

The method described herein is useful for undergraduate and high-school chemistry and environmental science lab courses. After this experiment, students will be able to make

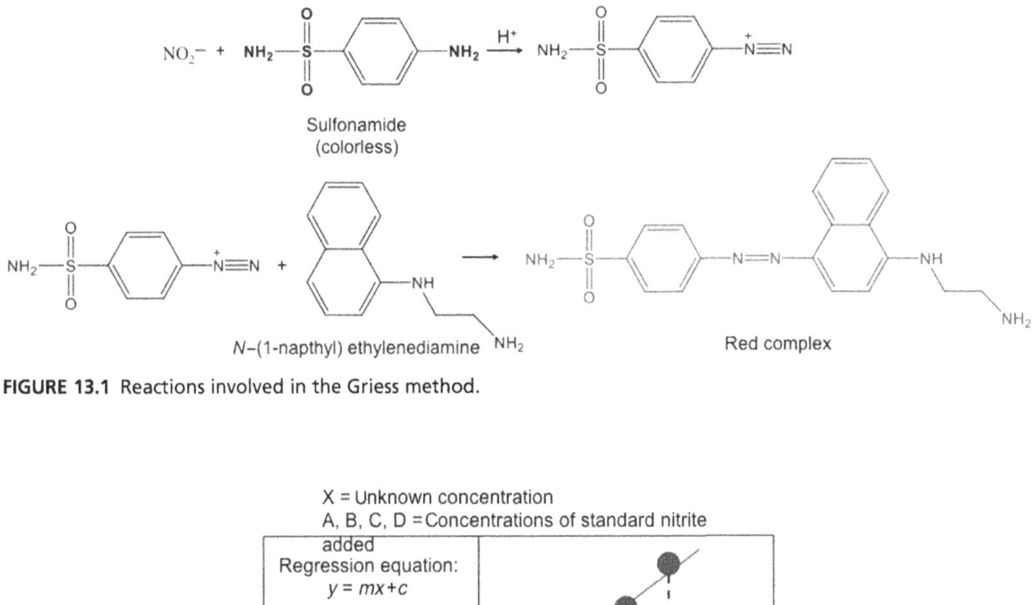

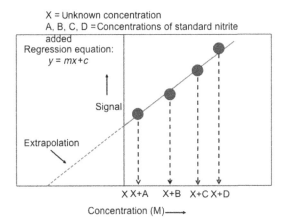

FIGURE 13.1 Reactions involved in the Griess method.

FIGURE 13.2 Example of calibration curve using standard addition method.

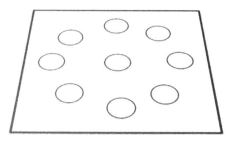

FIGURE 13.3 Design of the microfluidic device for nitrite detection.

μPAD using a permanent marker, perform a nitrite test, and use mobile phones as a tool for chemical analysis.

13.2 Microfluidic Device Design

The paper device consists of nine circular reaction zones. Each reaction zone is ~ 5 mm in diameter (Fig. 13.3).

13.3 Chemicals and Supplies

Deionized water, citric acid, sodium nitrite, methanol, *p*-amino benzenesulfonamide, *N*-(1-naphthyl)-ethylenediamine, and pond-water sample.

Whatman No. 1 paper, permanent marker, hot plate, clear packaging tape, micropipette and tips, centrifuge tubes, polystyrene template, petri dish, phone camera, and ImageJ software.

13.4 Hazards

p-Amino benzenesulfonamide may cause inflammation to the skin and eyes. Methanol is highly volatile, and may cause blindness or death if ingested in large amounts. It also has an anesthetic effect on the central nervous system and can cause serious erosion of the skin. Students are required to wear a long-sleeve lab coat, compatible chemical-resistant gloves, and chemical safety goggles.

13.5 Experimental Procedure

The number of students in this experiment is recommended to be 1–2. Details of the experimental steps are given in following sub-sections.

13.5.1 Sample Preparation

Solutions can be prepared while waiting for the paper device to dry to save time.

1. Obtain 1 mL of 25.0 mM sodium nitrite stock solution, unknown sample (UK1), and pond-water sample (UK2) from instructor.
2. Indicator solution: Can be prepared in a group of five students. Dissolve 0.86 g p-amino benzenesulfonamide, 0.19 g N-(1-naphthyl)ethylenediamine, and 6.34 g citric acid in 20% (v/v) water in methanol solution in a 100-mL volumetric flask and dilute to the mark. Store the indicator solution in an amber bottle.
3. Prepare 500 μL of each standard nitrite solution (e.g., 0.0780, 0.156, 0.312, 0.625, 1.25, and 2.50 mM) in centrifuge tubes from the stock solution using deionized water.
4. Filter the pond-water sample to remove suspended particles.
5. Repeat step 3 with filtered pond water instead of deionized water. We will use these solutions for the standard addition method.

13.5.2 Fabrication of Device

Following steps are needed to make paper device (Fig. 13.4).

1. Obtain two pieces of Whatman #1 filter paper (5 cm × 5 cm) and a plastic template. Be sure to handle the paper by the edges to avoid transferring finger oils to the paper's surface. We will make two devices: one for the regular method and another for the standard addition method.
2. Place the filter papers into a dry glass petri dish.
3. Pour fresh indicator solution into the dish to completely soak the papers and keep in dark.
4. Take out the presoaked filter papers from the indicator solution and keep in the dark for about 20 minutes to dry. Make sure the indicator solvent is completely evaporated before the next step.
5. Hand draw the microchannel pattern using a permanent marker and a plastic template. Apply the marker on both sides of the filter paper to make sure the hydrophobic circular boundaries of the reaction zones are leakproof. Once the solvent of the marker evaporates (in less than 5 minutes), put clear packaging tape on one side of the paper device.
6. Make sure the presoaked $\mu PADs$ are protected from light because the indicator is sensitive to light.

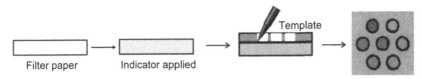

FIGURE 13.4 Steps for the fabrication of the paper device using a permanent marker for the determination of nitrite. A photo of the device after color development is given on right. *Reprinted with permission from Wang B, Lin Z, Wang M. Fabrication of a paper-based microfluidic device to readily determine nitrite ion concentration by simple colorimetric assay. J Chem Educ. 2015;92(4):733–736. Copyright (2015) American Chemical Society.*

13.5.3 Determination of Nitrite ion

Follow steps below to determine the concentration of nitrite.

1. Place the paper across the open end of a beaker or other container so that the device is suspended. This will prevent unwanted fluid flow due to contact with the benchtop.
2. Drop 1.5 µL of each standard solution, unknown 1 (UK1), pond-water sample (UK2), and blank (just DI water) to each of the circular reaction zones.
3. Put the paper device in the dark for 2 minutes to allow the reaction to develop.
4. Take a photo of the paper device with appropriate camera (digital camera or smartphone camera; whatever is available).
5. Determine the average gray values in the reaction zones using ImageJ software (*see* Appendix V) for both the regular and standard deviation methods.
6. Prepare a calibration curve that relates the concentration of the standards to the average gray value of standard solutions in the reaction zones and obtain the regression equation in Excel.
7. Determine the concentration of the analyte in the sample solution using the gray value of both unknown reaction zones and regression equation.
8. Also, estimate the detection limit, linearity, and error in your estimated concentrations of unknown samples and sensitivity for the paper-based method.
9. Determine the concentration of nitrite in unknown samples using the standard addition method. Compare your results.

13.5.4 Lab Report

Write a lab report following ACS style for journal articles. It should include an abstract, introduction, experimental, results and discussion, and conclusion sections with diagrams, tables, calculations, and images wherever necessary.

Consider the following points when writing the report:

1. Include a short description about the importance of measuring nitrite ions in environmental samples. Also describe colorimetric detection of nitrite ions using Griess method in paper device. Point out strengths and weakness of paper-based Griess method compared to conventional spectrophotometric method.
2. Outline the major steps and workflow involved in the experiment.
3. Add representative photos of the paper device and calibration curves.
4. Estimate the detection limit and concentration of unknowns. Also show representative calculations.
5. Compare the results from the regular method with the standard addition method and discuss any differences.

13.6 Additional Notes

1. The diazotization reaction between nitrite ions and the indicator performed on the prepared µ*PAD* should be protected from light.

2. The unknown sample solution (UK1) can be prepared by, for example, 100-fold dilution of a stock solution. Prepare at least two to three different sets of concentrations of unknown solution depending on the size of the class.
3. Spike the pond water with sodium nitrite solution such that the final concentration becomes 1 mM before giving it to students.

13.7 Assessment Questions

1. What are the key limitations of the paper-based method for nitrite determination?
2. In an experiment,[5] Sr content in tooth enamel was determined by the standard addition method. Solutions were prepared with a constant total volume of 10.0 mL containing 0.750 mg of dissolved tooth enamel. Variable concentrations of standard Sr were added in those solutions and signal was measured (see Table below).

Added amount (ng/mL)	Signal (AU)
1	28.0
2.5	34.3
5.0	42.8
7.5	51.5
10.0	58.6

   a. Find the concentration of Sr and its uncertainty in the 10-mL sample solution.
   b. Find the concentration of Sr in tooth enamel.

References

1. Bryan NS, Loscalzo J. In: Bendich A, ed. *Nitrite and Nitrate in Human Health and Disease*. New York: Springer Science; Business Media; 2011.
2. Table of regulated drinking water contaminants. United States Environmental Protection Agency; 2009. Available from: <https://www.epa.gov/sites/production/files/2016-06/documents/npwdr_complete_table.pdf>.
3. Wang B, Lin Z, Wang M. Fabrication of a paper-based microfluidic device to readily determine nitrite ion concentration by simple colorimetric assay. *J Chem Educ*. 2015;92(4):733–736.
4. Montgomery H, Dymock JF. Determination of nitrite in water. *Analyst*. 1961;86(102):414.
5. Porter VJ, Sanft PM, Dempich JC, Dettmer DD, Erickson AE, Dubauskie NA, et al. Elemental analysis of wisdom teeth by atomic spectroscopy using standard additions: An undergraduate instrumental analysis laboratory exercise. *J Chem Educ*. 2002;79(9):1114.

Colorimetric Determination of Multiple Metal Ions on μPAD

14.1 Background

Exposure to metal pollution through contaminated air and water can have a significant health impact to humans. For example: welding fumes are known to contain hazardous levels of particulate metals such as hexavalent chromium, nickel, copper, nitrous oxide, manganese, and lead.[1] Morbidity and mortality from occupational respiratory diseases are estimated to cost ten billion dollars each year in the United States. The health effects associated with exposure to airborne metals are extensive and include cancers of the respiratory, hematopoietic, renal, nervous, hepatic, and digestive systems, immune disorders, birth and developmental defects, neuropathy, nephropathy, and specific respiratory diseases such as asthma, bronchitis, chronic obstructive pulmonary disease (COPD), and emphysema.[2]

In this laboratory experiment,[3,4] we are going to simultaneously measure the concentration of four metal ions using a colorimetric $\mu PADs$ method. The metal ions we are going to measure are Fe, Ni, Cu, and Cr. This method allows the detection of an individual metal in the presence of possible interfering metals. This μPAD approach is different from other paper-device experiments given in this book because it includes sample pretreatment on the device as well as addition of stabilizing agents to give the device long-term shelf life. The absence or presence of these metals will be confirmed by visual comparison of color intensity. For quantitative analysis, a photo of the reaction color will be taken and will be processed using ImageJ software. Individual metal-calibration curves will be generated and concentration of the metals in unknown sample will be measured. We will use a flatbed scanner instead of a phone camera for detection to get a high-resolution and well-focused image. This detection method is typically unaffected by external lighting conditions.

14.1.1 Colorimetric Assays

We will determine the Fe content using a popularly used phenanthroline method.[5] In this method the red/orange ferroin complex formed by the reaction of Fe(II) with 1,10-phenanthroline is detected (Fig. 14.1).

As this reaction requires the iron in ferrous state, it is necessary to convert any ferric ions present in the sample to ferrous before the color formation reaction takes place. We will use hydroxylamine as a mild reducing agent that reduces Fe^{3+} to Fe^{2+} (see reaction 14.1).

$$2Fe^{3+} + 2NH_2OH \cdot HCl + 2OH^- \leftrightarrow 2Fe^{2+} + N_2 + 4H_2O + H^+ + Cl^- \tag{14.1}$$

Laboratory Methods in Microfluidics. DOI: http://dx.doi.org/10.1016/B978-0-12-813235-7.00014-3

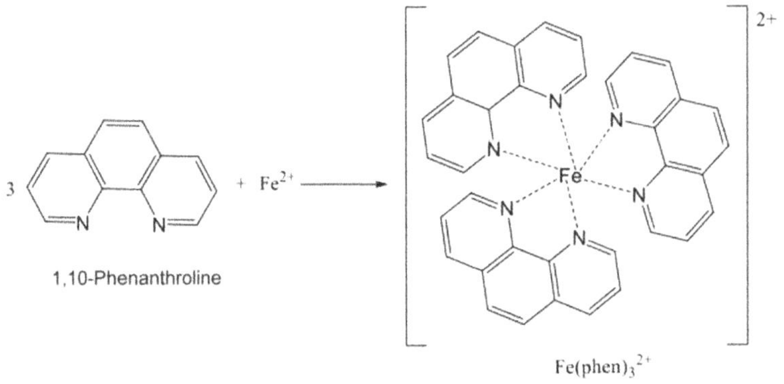

FIGURE 14.1 Reaction of ferrous ion with phenanthroline.

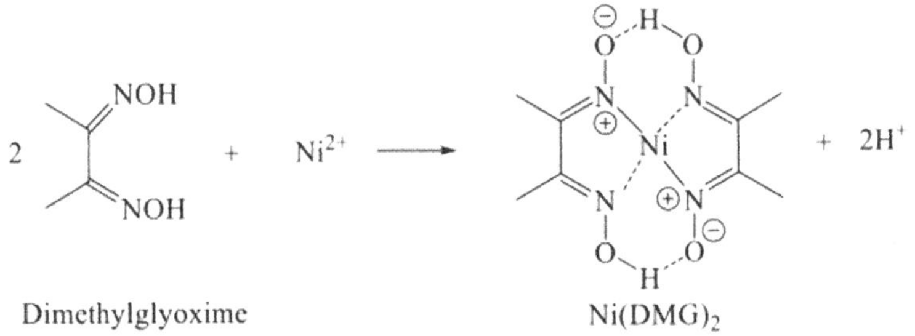

FIGURE 14.2 Reaction of nickel ion with DMG.

In addition, hydroxylamine also acts as a masking agent by complexing possible interfering metals such as Ni, Zn, Cd, and Co in the pretreatment zone. We will also apply poly (acrylic acid) in the detection zone to reduce the mobility of $[Fe(o\text{-}phen)_3]^{2+}$ in paper. Poly (acrylic acid) has low mobility in paper, and its negative charges serve to keep the positively charged complexation product from moving once formed.[4]

Ni will be determined using dimethylglyoxime (DMG).[6] DMG produces a bright pink Ni(II) complex (Fig. 14.2).

Sodium fluoride (0.5 M) and acetic acid (10 mM) are used to mask interference from Fe, Cu, and Co. An ammonium hydroxide solution (pH 9.5) at the detection zone is effective at keeping the optimal pH of ~9 for the assay.

The concentration of copper ions will be determined by bathocuproine (BC).[7,8] The disodium salt of BC disulphonic acid forms an orange complex with Cu^+ (Fig. 14.3). This method is chosen because the reagent does not complex Fe and is the most sensitive of all the cuproines commonly used for Cu detection. Hydroxylamine is added to the pretreatment and detection zones for reduction of Cu^{2+} to Cu^+. To stabilize the orange complex, an acetic acid/NaCl buffer of pH 4.5 is used to set the pH at the detection zone. The Cl^- ion stabilizes the orange $Cu(BC)_2$ complex. This issue is overcome by adding polyethylene

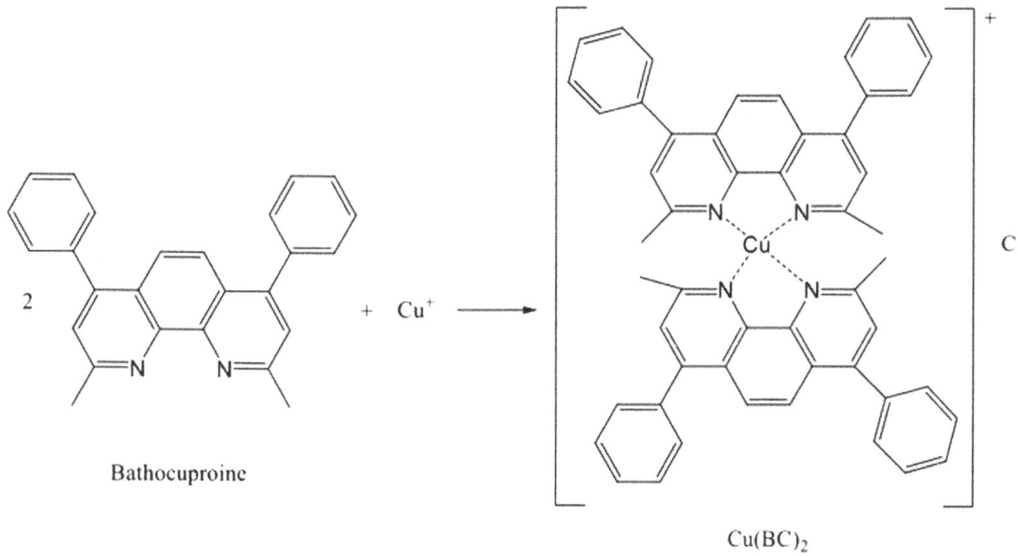

FIGURE 14.3 Reaction of Cu$^+$ with BC.

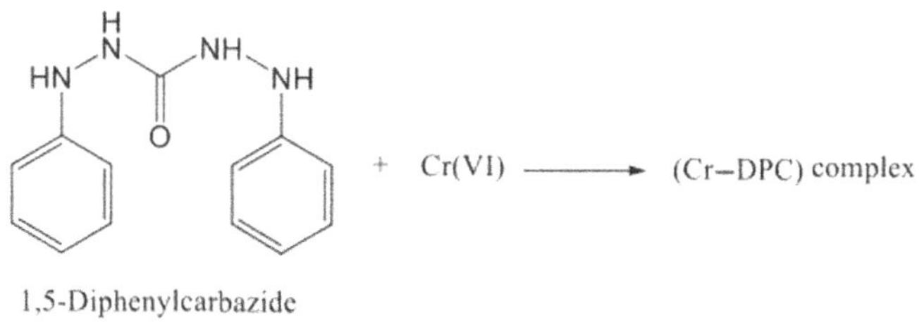

FIGURE 14.4 Reaction used for determination of chromium.

glycol (PEG) to the BC solution. The presence of dried PEG on the device helps to reduce the hydrophobicity in the detection reservoir caused by the BC and allows the sample to flow into the reservoir[4].

The total Cr is determined using tetravalent cerium Ce(IV) and 1,5-diphenylcarbazide (1,5-DPC) as oxidizing and colorimetric reagents, respectively.[9] Tetravalent cerium oxidizes all forms of soluble Cr to Cr(VI) for reaction with 1,5-DPC. 1,5-DPC has been used as a selective Cr(VI) reagent for decades. 1,5-DPC reduces Cr(VI) to Cr(III) and is itself oxidized to diphenylcarbazone (DPCO). DPCO complexes with the generated Cr(III) to form an intense purple-colored complex (Fig. 14.4). Phthalic anhydride stabilizes 1,5-DPC on *μPADs*. PDDA is added to stabilize the reaction product between Cr and 1,5-DPC and to prevent the complex from flowing to the edges of the hydrophilic channels.

An incineration ash sample, certified for the metals Ag, Al, Ba, Cd, Cr, Cu, Fe, Mg, Mn, Ni, Pb, Zn, Co, and V, will be used to simulate a real-world particulate metal hazard.

14.2 Microfluidic Device Design

The μPAD design[4] (Fig. 14.5) includes a central sample reservoir (10 mm) surrounded by four channels that lead to detection reservoirs (2 mm). The channels also contain pretreatment zones (2 × 3 mm) that can be used for the addition of reagents needed to either oxidize/reduce metals or to complex metals that interfere with a specific assay. An elliptical shape is used for this zone so that sufficient volumes of reagents can be added without leaking into the detection or sample reservoirs. The pretreatment zone is especially useful when reagent solutions has a drastically different pH than that needed for the formation of the final colored metal complex. By impregnating the pretreatment zones with masking agents, interfering metals are complexed. Moreover, analytes are reduced to the appropriate oxidation state before the sample reaches the detection reservoirs by adding reductants in the pretreatment zones. Control reservoirs present in the device help to distinguish any colored compounds from the real products.

14.3 Chemicals and Supplies

Deionized (DI) water, acetone, methanol, chloroform, glacial acetic acid, iron(III) chloride hexahydrate, nickel(II) sulfate hexahydrate, copper(II) sulfate pentahydrate, ammonium dichromate (VI), 1,5-DPC, phthalic anhydride, 1,10-phenanthroline, hydroxylamine, DMG, BC, PEG, sodium fluoride, cerium(IV) ammonium nitrate, polydiallyldimethylammonium chloride (PDDA, medium molecular weight), ammonium hydroxide, sodium acetate trihydrate, and metal-containing certified industrial incineration ash samples.

Whatman No. 1 filter paper, wax printer, and desktop flatbed scanner.

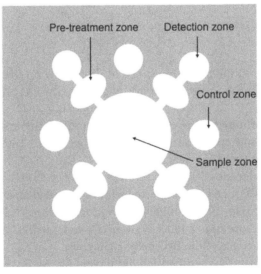

FIGURE 14.5 Design of the microfluidic device used in this experiment.

14.4 Hazards

Solutions should be prepared under hood. Personal protective equipment is recommended during the experiment.

14.5 Experimental Procedure

Recommended size: group of two.

14.5.1 Fabrication of the Device

1. Design the paper device in CorelDraw or Adobe Illustrator.
2. Print the design onto the filter paper using a wax printer.
3. Melt the wax into the paper to create hydrophobic channel by placing the paper onto a 150°C hot plate for 120 seconds.
4. Let the paper cool. Apply packing tape to one side of the filter paper to prevent reagents from leaking through the device. At least six devices are needed for this experiment.

14.5.2 Preparation of Solutions

Obtain Fe, Ni, Cu, and Cr standard solutions at 1000 ppm and prepare the following solutions. Reagents need to be made fresh each time.

1. A 10 mM acetate buffer of pH 4.5.
2. *For Cr detection*: A solution containing 15 mg/mL of 1,5-DPC and 40 mg/mL of phthalic anhydride in acetone; 0.35 mM of ceric(IV) ammonium nitrate; 5% (w/v) PDDA.
3. *For Fe*: 8 mg/mL of 1,10-phenanthroline in acetate buffer; 0.1 g/mL of hydroxylamine in acetate buffer; 0.7 mg/mL of poly(acrylic acid);
4. *For Ni*: 0.5 M of NaF; 60 mM solution of DMG in methanol;
5. *For Cu*: 50 mg/mL BC with 40 mg/mL PEG 400 in chloroform.

14.5.3 Determination of Metal Ions

1. *Cr determination*: Add 0.5 μL ceric(IV) ammonium nitrate (0.35 mM) to the pretreatment zone twice. Add 0.5 μL of PDDA (5% w/v) in the detection zone. Deposit a 0.5 mixture of μL of 15 mg/ml 1,5-DPC, and 40 mg/mL phthalic anhydride on the detection zone.
2. *Fe determination*: Add 1 μL of hydroxylamine solution to the detection reservoir. After this solution dries, add 0.35 μL of poly(acrylic acid) (0.7 mg/mL) to the detection reservoir followed by two 0.4 μL aliquots of 1,10-phenanthroline also in the detection reservoir. Allow the device to dry completely between each addition of reagent.
3. *Ni determination*: Add two 0.4 μL aliquots of NaF (0.5 M) followed by 0.4 μL of acetate buffer (pH 4.5) to the channel reservoir. Add five 0.4 μL aliquots of DMG to the detection reservoir followed by two 0.4 μL aliquots of ammonium hydroxide (pH 9.5). Allow the device to dry completely between each addition of reagent.

4. *Cu determination*: Add 1 µL of hydroxylamine (0.1 g/mL) followed by 0.5 µL of acetate buffer (10 mM, pH 4.5) to the detection reservoir. Then add two 0.5 µL aliquots of the BC/PEG solution to the detection reservoir. Allow the device to dry completely between each addition of reagent.
5. Add 10 µL of sample containing a mixture of four metal ions. Allow the sample wick up to the detection and observe the change in color.
6. Scan the device using a desktop flatbed scanner and store the image in JPEG format.
7. Using ImageJ software, obtain gray values in each of the detection reservoirs, plot calibration curves for each metal ion, and then determine the concentration of unknown mixture (*see* Appendix V for details on using ImageJ software).
8. Obtain the calibration curve for all metal ions.
9. Determine the detection limit, linear range, relative standard deviation, and quantification range for each metal ion. The background values were used to determine the baseline intensity for detection-limit calculations.

14.5.4 Lab Report

Write a lab report following ACS style for journal articles. It should include an abstract, introduction, experimental, results and discussion, and conclusion sections with diagrams, tables, calculations, and images wherever necessary.

Consider the following points when writing the report:

1. Include a short description of the importance of quantitation of metal ions in environmental samples and colorimetric methods for the metal ions in paper device.
2. Outline the major steps and workflow involved in the experiment.
3. Add representative photos of the device and calibration curves.
4. Include results of unknown sample with error and show calculations.

14.6 Additional Notes

1. While making the paper device, the wax heating temperature and time may vary based on the type of paper and wax used. It is better to optimize these parameters.
2. While only detection of Fe, Ni, Cu, and Cr is discussed here, the device can easily be modified to accommodate detection of other metals. Also, individual assays can often be optimized to eliminate specific interferences that are known to exist at high levels in a particular work environment.

14.7 Assessment Questions

1. What are the roles of hydroxylamine used in iron assay?
2. Why is it necessary to use sodium fluoride, acetic acid, and ammonium hydroxide in nickel assay?
3. Describe the role of polyethylene glycol in copper assay and explain how this works.

References

1. Quansah R, Jaakkola JJ. Paternal and maternal exposure to welding fumes and metal dusts or fumes and adverse pregnancy outcomes. *Int Arch Occup Environ Health.* 2009;82(4):529−537.

2. Nelson DI, Concha-Barrientos M, Driscoll T, Steenland K, Fingerhut M, Punnett L, et al. The global burden of selected occupational diseases and injury risks: methodology and summary. *Am J Ind Med.* 2005; 48(6):400−418.

3. Cate DM, Nanthasurasak P, Riwkulkajorn P, L'Orange C, Henry CS, Volckens J. Rapid detection of transition metals in welding fumes using paper-based analytical devices. *Ann Occup Hygiene.* 2014; 58(4):413−423.

4. Mentele MM, Cunningham J, Koehler K, Volckens J, Henry CS. Microfluidic paper-based analytical device for particulate metals. *Anal Chem.* 2012;84(10):4474−4480.

5. Saywell LG, Cunningham BB. Determination of iron: colorimetric *o*-phenanthroline method. *Ind Eng Chem Anal Ed.* 1937;9(2):67−69.

6. Mitchell AM, Mellon MG. Colorimetric determination of nickel with dimethylglyoxime. *Ind Eng Chem Anal Ed.* 1945;17(6):380−382.

7. Smith GF, Wilkins DH. New colorimetric reagent specific for copper. *Anal Chem.* 1953;25(3):510−511.

8. Borchardt LG, Butler JP. Determination of trace amounts of copper. *Anal Chem.* 1957;29(3):414−419.

9. Farag AB, El-Wakil AM, El-Shahawi MS. Qualitative and semi-quantitative determination of chromium (VI) in aqueous solution using 1,5-diphenylcarbazide-loaded foam. *Analyst.* 1981;106(1264):809−812.

Analysis of a Mixture of Paracetamol and 4-Aminophenol in a Paper-Based Microfluidic Device

15.1 Background

Paracetamol (PA) is a widely used medicine. It is a mild painkiller (analgesic) and reduces the temperature of patients with fever (antipyretic). This drug was discovered in mid-1940s, and today there are hundreds of common products containing PA. However, there are concerns about the health effects of this drug such as risk of liver injury and allergic reactions. The industrial synthesis of PA is carried out by the acetylation of 4-aminophenol (4-AP) with acetic anhydride. Therefore *p*-aminophenol is the main impurity of PA-containing pharmaceuticals, which can be due to the degradation of PA or by the presence of the reagent in the final product. It is reported that 4-AP has nephrotoxic and teratogenic properties in humans.[1] Therefore measurement of both PA and *p*-aminophenol in PA-containing medicine is very important (Fig. 15.1).

In this experiment, which is based on reference 2, we combine a paper-based microfluidic device and electrochemical detection system for the simultaneous determination of PA and 4-AP. The mixture of these chemicals is separated and detected in the same paper device.

This experiment utilizes electrochemical detection instead of colorimetric detection as used in previous experiments. Electrochemical detection has also been widely combined with paper analytical devices because of the possibility of miniaturization, portability, low cost, and high sensitivity.[3] Amperometry is one of the electrochemical methods in which current or change in current is measured in the presence of given chemical species. We will fabricate a three-electrode electrochemical cell system on the paper device. The electrochemical system consists of a pseudoreference electrode near the working electrode to minimize the ohmic drop effects. Whatman No. 1 chromatographic paper is one of the most employed substrates for the construction of paper-based analytical devices. However, this type of paper is not suitable for the separation of mixture containing PA and 4-AP. The two analytes have very different pKa values, 9.8 for R−OH group of PA and 5.3 for the R−NH^{3+} group of 4-AP. We will therefore use cation-exchanger paper (Whatman P81) to facilitate the separation process.[2] The P81 is a strong cation exchanger of high capacity with an ion-exchange capacity of 18.0 μeq/cm^2. The ester-linked orthophosphoric acid group with Na$^+$ counter ions is responsible for the ion-exchange process. Thus, when protonated 4-AP is in the paper-based column there is an exchange process between sodium ions and 4-AP. This ion-exchange process

Laboratory Methods in Microfluidics. DOI: http://dx.doi.org/10.1016/B978-0-12-813235-7.00015-5

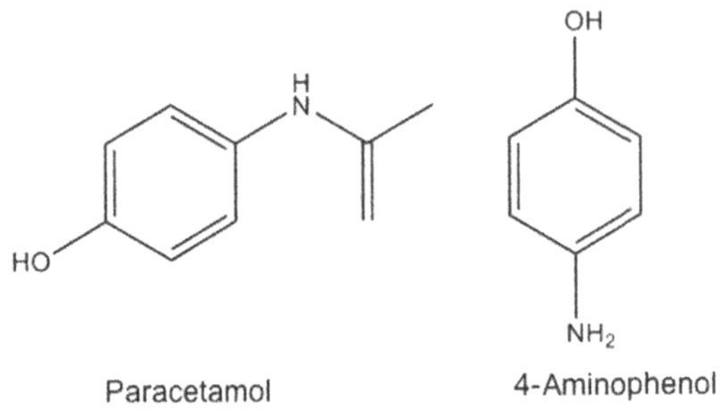

FIGURE 15.1 Structures of paracetamol and 4-aminophenol.

retains 4-AP molecules on the paper-based column. The retention time of 4-AP is higher because it interacts with the negatively charged functional groups of the paper. The degree of separation of 4-AP from PA can be described by the resolution (R_s), measured as the difference in retention time (t_R) of the two analytes divided by their average peak width (w_b).

In this experiment, the eluent and samples do not need to be filtered before injection in the paper-based column as in traditional methods because the paper itself filters particles. In addition, this method does not require an external pump to move the sample and/or apply high voltage as in electrophoretic separation to separate the chemical species. Instead the samples are moved from one place to other by capillary action.

This experiment can be completed in a 3−4 hour laboratory session and is appropriate for senior level undergraduate courses in chemistry, and pharmaceutical science. After the completion of this experiment, students will learn about the fabrication of paper device, microfabrication of metal electrodes, and separation of organic compounds, using electrochemical detection system.

15.2 Microfluidic Device Design

The paper device is ~50 mm long with a hydrophilic separation channel of 2 mm wide. The geometric areas of the electrodes are 1.0 mm² for the working electrode and 2.0 mm² for the counter and reference electrodes. The eluent enters the device through region A and is directed to the detection system by the hydrophobic walls. The analytes are separated in region B and are detected in region C. The absorbent pad area-D is made by pressing multiple sheets of filter paper. The absorbent pad allows that the previously detected analytes are directed to this waste zone, without impairing subsequent analysis. Thus, the absorbent pad works as a waste reservoir for the disposal of eluent and samples (Fig. 15.2).

15.3 Chemicals and Supplies

PA, acetic acid, sodium acetate, sodium hydroxide, 4-AP, methanol, MilliQ water, and commercial PA-containing drug tablets.

Whatman chromatographic P81 paper, wax printer, sputtering system, hot plate, pH meter, potentiostat, micropipette, and tips.

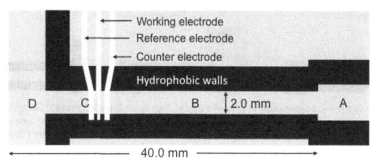

FIGURE 15.2 Design of the electrochemical paper microfluidic device. *Reprinted from Shiroma LY, Santhiago M, Gobbi AL, Kubota LT. Separation and electrochemical detection of paracetamol and 4-aminophenol in a paper-based microfluidic device. Anal Chim Acta. 2012;725:44–50. Copyright (2012) with permission from Elsevier.*

15.4 Hazards

Eye protection and other appropriate protective equipment should be worn during the experiment.

15.5 Experimental Procedure

15.5.1 Preparation of Solutions

Prepare the following solutions. Estimate the amount of solution needed.

1. M acetate buffer of pH 4.5.
2. Working standard solutions (0.05–2.0 mM) of PA and 4-AP in 0.1 M acetate buffer solution.
3. Dissolve the capsule in 1.00 L volumetric flask with 25.0 mL 0.1 M acetate buffer solution.

15.5.2 Fabrication of the Device

Follow the steps given below to make the device (Fig. 15.3).

1. Design the chip layout in drawing software like Microsoft PowerPoint.
2. Cut the chromatographic papers into A4 size.
3. Print the hydrophobic lines to define a hydrophilic channel on Whatman P81 chromatographic paper using a wax printer.
4. Place the printed paper on a hot plate set at 110°C for 120 seconds.
5. Remove the patterned paper from the hot plate and allow it to cool to room temperature.
6. Cut each channel individually for the construction of the electrodes.
7. Position a physical mask made of copper at the end of the channel and deposit a thin film of gold (200 nm) on the mask using a sputtering system.
8. Remove the mask, and leave the gold tracks marked on the paper.

FIGURE 15.3 Steps in the microfabrication of gold electrodes. *Reprinted from Shiroma LY, Santhiago M, Gobbi AL, Kubota LT. Separation and electrochemical detection of paracetamol and 4-aminophenol in a paper-based microfluidic device. Anal Chim Acta. 2012;725:44–50. Copyright (2012) with permission from Elsevier.*

15.5.3 Electrochemical Measurements

1. Carry out all amperometric measurements in 0.1 M acetate buffer solution.
2. Fix the potential of the working electrode at 0.4 V vs pseudo Au reference electrode.
3. Drop 0.5 μL of the sample at 12 mm from the working electrode.
4. Record the chromatogram and use the peak area to analyze the results.
5. Normalize the chromatograms by the injected time ($t = 0$ seconds).
6. Obtain calibration curves for both PA and 4-AP using the area under the curve as the signal and then regression equation.
7. Estimate the amount of the analyte in the sample. Estimate the detection limits and coefficient of variance using 0.8 mM of mixture.

15.5.4 Lab Report

Write a lab report following ACS style for journal articles. It should include an abstract, introduction, experimental, results and discussion, and conclusion sections with diagrams, tables, calculations, and images wherever necessary.

Consider the following points when writing the report:

1. Include a short description of the importance of quantitation of PA and its degradation product and electrochemical methods for the detection of these chemical species.
2. Outline the major steps and workflow involved in the experiment.
3. Add representative photos of the device, chromatograms, and calibration curves.
4. Include the results of the unknown sample and errors including calculations.

15.6 Additional Notes

1. The paper electrochemical device can be used for multiple separations (more than 50) and should be stable for at least 3 days.
2. The fabrication of gold electrodes on paper substrates can also be done by direct inkjet printing of gold electrodes.

15.7 Assessment Question

1. What are the advantages of using a paper electrochemical system compared to spectrophotometric determination of PA and 4-AP?

References

1. Ellis F. In: Osborne C, Pack M, eds. *Paracetamol: A Curriculum Resource*. 1st ed. London, UK: Royal Society of Chemistry; 2002.

2. Shiroma LY, Santhiago M, Gobbi AL, Kubota LT. Separation and electrochemical detection of paracetamol and 4-aminophenol in a paper-based microfluidic device. *Anal Chim Acta*. 2012;725:44–50.

3. Mettakoonpitak J, Boehle K, Nantaphol S, Teengam P, Adkins JA, Srisa Art M, et al. Electrochemistry on paper-based analytical devices: a review. *Electroanalysis*. 2016;28(7):1420–1436.

16

Synthesis of Gold Nanoparticles on Microchip

16.1 Background

Nanoscience studies the synthesis, manipulation, and use of nanomaterials. Nanomaterials are particles that display feature sizes of <100 nm in at least one dimension. In bulk at macroscale, gold is gold in color, but in nanoscale particles, gold is red to purple in color. Nanoparticles (NPs) differ from bulk materials because of dimensional confinement of electron density in the NP, which leads to local surface plasmon resonance (LSPR). NPs of different sizes and shapes display different plasmon resonances and as a consequence absorb light of different wavelengths, thereby giving rise to different colors of NP dispersions that can be monitored by UV−vis spectroscopy.[1] NP research has been an intense subject of scientific focus in the past several decades because of its potential applications in biomedical, optical, and electronic fields.

There are several methods for the synthesis of NPs. Among them, the synthesis of NPs by reduction of metal salts is the most widely used method. Gold nanoparticles (AuNPs) are synthesized by the reduction of chloroauric acid ($HAuCl_4$, hydrogen tetrachloroaurate). Sodium citrate solution as a reducing agent is added to a stirring hot aqueous solution of $HAuCl_4$. This reaction reduces Au^{3+} ions to neutral gold atoms. As the gold atoms are gradually formed, they accumulate to form AuNPs making the solution colloidal (Fig. 16.1).[2] Depending on particle size and shape, AuNPs display a surface plasmon resonance peak (λ_{max}) between 500 nm and 600 nm, which gives rise to the pink to purple color. In this method, the citrate ions not only act as reducing agent but also as a capping agent to prevent aggregation of NPs.

Upon introducing the sodium citrate to the boiling Au(III) chloride solution, sodium citrate is turned to citric acid. The citric acid then reduces Au^{3+} to Au^0 (atomic gold). Then, atomic gold aggregates to form AuNPs. The $Au^{3+}_{(aq)}$ solution is initially yellow in color. Upon addition of sodium citrate, the solution suddenly becomes colorless, turns blue within a fraction of a second, and then slowly converts to wine-red NPs. Varying the concentration of sodium citrate can control the size of NP. More citrate in the system produces smaller AuNPs and vice versa. A portion of the sodium citrate is used to reduce $Au^{3+}_{(aq)}$ and the remaining sodium citrate ions are available for stabilizing the particles. NP aggregation occurs until the total surface area of all particles becomes small enough to be covered by existing citrate ions. Therefore less sodium citrate in the medium will cause the small particles to aggregate into bigger ones.

NPs are traditionally synthesized using batch techniques. Recently, interest in the continuous flow production of NPs has increased. The flow-reactor method offers several

Laboratory Methods in Microfluidics. DOI: http://dx.doi.org/10.1016/B978-0-12-813235-7.00016-7

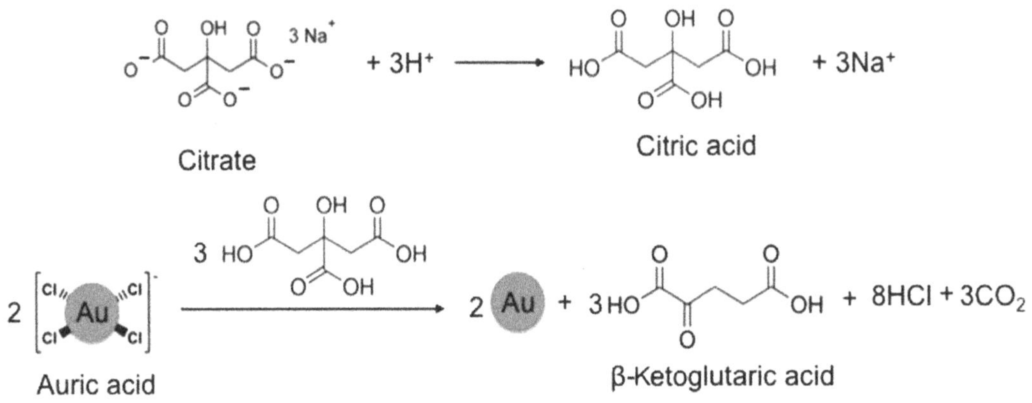

FIGURE 16.1 Reactions involved in gold nanoparticle synthesis using citrate as reducing agent.

advantages over the batch method. The microreactor has high surface-area-to-volume ratios significantly increasing the interfacial contact area of reactants in the confined narrow channels. The microreactor flow synthesis is faster, offers better heat transfer, and reduces hazardous waste, making it safe for the synthesis of hazardous compounds and isolation of compounds sensitive to air and/or moisture. The small reaction volumes in microreactors allow for efficient heat transfer. Using the flow reactor, a variety of factors, such as temperature, residence time, and ratio of reactants can be manipulated to have NPs of desired sizes and shapes.[3-5]

The formation of NP can be observed by the naked eye as color change. The size of the NP can also be determined by using UV–vis spectroscopy in addition to the use of electron microscopes. In this process, the wavelength for maximum absorption is measured and the size is calculated according to:[6]

$$d = \ln\left(\frac{\lambda_{max} - 512}{6.53}\right) * 46.29 \qquad (16.1)$$

where d is the NP diameter and λ_{max} is the SPR wavelength of the nanoparticle solution. However, this equation holds good for particles bigger than 25 nm.

In this experiment,[7,8] we will use a cost-effective flow-based microchip method for the synthesis of AuNPs. The microchip is placed on a hot plate. Solutions of $HAuCl_4$ and sodium citrate are then injected into the microchannel at a given flow rate. The reactant solutions can be injected manually or automated syringe pumps can be used for better and precise injection (or using a peristaltic pump as a cheaper alternative). The resulting citrate-capped AuNPs are collected from the outlet. A schematic of microchip NP synthesis setup schematic is shown in Fig. 16.2.

This experiment can be completed in a 3-hour lab session and is suitable for upper-division chemistry, nanotechnology, and material science courses. Even freshmen students as individuals or as a group can work on this experiment. This application of microfluidic

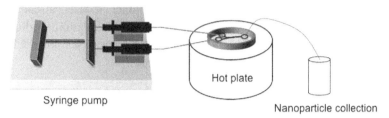

FIGURE 16.2 Schematic of microchip nanoparticles synthesis setup.

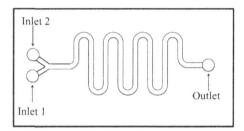

FIGURE 16.3 Design of the nanoparticle synthesis chip.

system integrates important concepts in chemical engineering, inorganic synthesis, nanomaterial science, and analytical chemistry. Students will be able to learn flow synthesis of metal NP, stabilization of NPs, concept of surface plasmon resonance, and the determination of NP size using UV–vis spectrophotometry.

16.2 Microfluidic Device Design

The microfluidic device consists of two inlets, one for $HAuCl_4$ solution and another for sodium citrate solution, and one outlet for receiving the NPs. The microchip (either a PDMS chip or a glass chip) can be made by students in the lab or purchased from commercial vendors (*see* Appendix VI for vendor list) (Fig. 16.3).

16.3 Chemicals and Supplies

Deionized (DI) water, NaOH, $HAuCl_4$, sodium citrate, and methanol.

Microchip, syringe pump, 1-mL syringes, tubing, hot plate, UV–vis spectrophotometer, and electron microscope (optional).

16.4 Hazards

Always wear safety goggles when working in the lab. Use of lab gloves is strongly recommended throughout the experiment. Chloroauric acid trihydrate is corrosive and hygroscopic. It can cause eye and skin burns upon contact. It also causes gastrointestinal tract burns if swallowed and burns the respiratory tract by inhalation. Sodium citrate may cause irritation to skin, eyes, and respiratory tract. All lab exercises should be carried out under the supervision of trained and qualified personnel.

16.5 Experimental Procedure

16.5.1 Solution Preparation

Prepare following solutions.

1. Prepare 1 mM $HAuCl_4$ and 1% sodium citrate solution.

16.5.2 Synthesis of Nanoparticle

Follow the steps given below to synthesize gold nanoparticles.

1. Set the hot plate to the desired temperature. The recommended temperature is 60°C but temperature can be varied to study its effect on NP synthesis.
2. Fill a 1-mL syringe with $HAuCL_4$ solution and a second syringe with sodium citrate solution. We encourage students to vary the concentration of these two solutions. A higher concentration of Au^{3+} solution can be used for enhanced visual observation of the colors of the AuNP.
3. Attach two syringes to the syringe pump where the plungers are aligned. Remove any excess solution from the ends if the syringe needs to be adjusted. Then, attach the syringes to the tubing leading to the channel inlets.
4. Place the outlet tubing into a cuvette.
5. Select a flow rate and start the syringe pump. We will run a total of three trials with [gold]:[citrate] ratio of 1:1, 1:2, and 1:3. To get this ratio, set the flow rate of gold solution and citrate solution accordingly on syringe pump for.
6. Take an absorption spectrum in the visible range of each product and record the λ_{max} and peak width at half maximum displayed.
7. Estimate the size of your AuNPs at various conditions using Eq. (16.1).

16.5.3 Lab Report

Write a lab report following ACS style for journal articles. It should include an abstract, introduction, experimental, results and discussion, and conclusion sections with diagrams, tables, calculations, and images wherever necessary.

Consider the following points when writing the report:

1. Include a short description of flow-based AuNP synthesis including its strengths and weakness.
2. Outline the major steps involved in the experiment.

3. Add representative spectra of the NP solution.

4. Estimate the size of NPs at various conditions. Show representative calculations.

16.6 Additional Notes

1. Manual injection of reactants is possible but it is difficult to ensure equal pressure and even flow rate by manual injection. Therefore automated pumps are strongly recommended. Syringe pumps are the easiest one to use and attach to the device with reduced numbers of junctions to ensure minimal leaking. However, if the price of the pump is a concern, a peristaltic pump is a good alternative.

2. Other magnetic particles such as iron particles can be synthesized following a similar procedure.

16.7 Assessment Questions

1. What are the three advantages and disadvantages of performing synthesis in a microfluidic device in flow condition?

2. How were the NPs prevented from aggregating together in this experiment? Explain.

3. What flow rate did you use in your experiment? Were there any visible differences in products? How do you expect the flow rate to affect the reaction?

References

1. Binns C. *Introduction to Nanoscience & Nanotechnology*. Florida, USA: John Wiley & Sons, Inc.; 2010.

2. Fedlheim DL, Foss CA. *Metal Nanoparticles: Synthesis, Characterization, and Applications*. 1st ed. New York: CRC Press; 2001.

3. Zhao CX, He L, Qiao SZ, Middelberg AP. Nanoparticle synthesis in microreactors. *Chem Eng Sci.* 2011;66(7): 1463–1479.

4. Marre S, Jensen KF. Synthesis of micro and nanostructures in microfluidic systems. *Chem Soc Rev.* 2010;39(3):1183–1202.

5. Rahman MT, Rebrov EV. Microreactors for gold nanoparticles synthesis: from Faraday to flow. *Processes.* 2014;2(2):466–493.

6. Haiss W, Thanh NT, Aveyard J, Fernig DG. Determination of size and concentration of gold nanoparticles from UV–vis spectra. *Anal Chem.* 2007;79(11):4215–4221.

7. Feng ZV, Edelman KR, Swanson BP. Student-fabricated microfluidic devices as flow reactors for organic and inorganic synthesis. *J Chem Educ.* 2015.

8. Ftouni J, Penhoat M, Addad A, Payen E, Rolando C, Girardon JS. Highly controlled synthesis of nanometric gold particles by citrate reduction using the short mixing, heating and quenching times achievable in a microfluidic device. *Nanoscale.* 2012;4(15):4450–4454.

17

Flow Synthesis of Organic Dye on Microchip

17.1 Background

The synthetic azo dyes with $-N=N-$ structure (called an azo group) are widely used for various applications. Most azo dyes contain only one azo group, but others may contain two, three or more. The azo group joins together aromatic rings, resulting in an extended system of conjugated multiple bonds. The conjugated system is characteristic of all organic dyes. Colored compounds have such extended conjugated systems that their "UV" absorptions extend into the visible region. For a substance to have color, it must absorb within the 400−700 nm region of the spectrum. The nature of the aromatic substituents on both sides of the azo group controls the colors as well as the water solubility of the dyes. Azo dye compounds come in a broad range of colors, including yellows, oranges, reds, browns, and blues.[1]

Azo dyes are prepared in a two-step reaction (Fig. 17.1).[2] The first step involves the reaction of aromatic amine with nitrous acid to give a diazonium salt. The formation of diazonium intermediate is called diazotization. For example: aniline reacts with dilute hydrochloric acid to form anilinium chloride. This chloride forms a diazonium salt when reacted with nitrous acid. Aryl diazonium salts are stable for a reasonable period of time if kept in an aqueous solution at 0−5°. The energy requirements for the reaction are low, since most of the chemistry occurs at or below room temperature. The diazonium ion is an electron-deficient (electrophilic) intermediate. An aromatic compound attached to the diazonium, suitably rich in electrons (nucleophilic), makes the diazonium even more electrophilic. The most commonly used nucleophilic species are aromatic amines and phenols[3].

In the second step, the diazonium salt reacts with a coupling compound such as a phenol and gives an azo dye. The addition of phenol to the diazonium ion is called the diazonium-coupling reaction.

In this experiment, adapted from references 4,5, students will synthesize an azo dye in a flow-based microfluidic platform. The microfluidic chip is placed on an ice-bath cooled petri dish with an aniline solution. Then sodium nitrite and phenol solutions are injected from two inlets. The resulting dye is acidified until it crashes out of solution. The product is analyzed with IR-absorption spectroscopy and UV−vis absorption spectroscopy.

The color of the azo dye product allows for a quick and easy initial assessment of the success of the reactions. The flow-based synthesis reaction offers three advantages[6]: (1) simply modifying the chip design can perform multistep synthesis; (2) efficiency in heat transfer is demonstrated in the cooling condition with an ice bath. The small reactant volume again makes rapid cooling possible; and (3) we improve the safety of the reaction by merging the

Laboratory Methods in Microfluidics. DOI: http://dx.doi.org/10.1016/B978-0-12-813235-7.00017-9

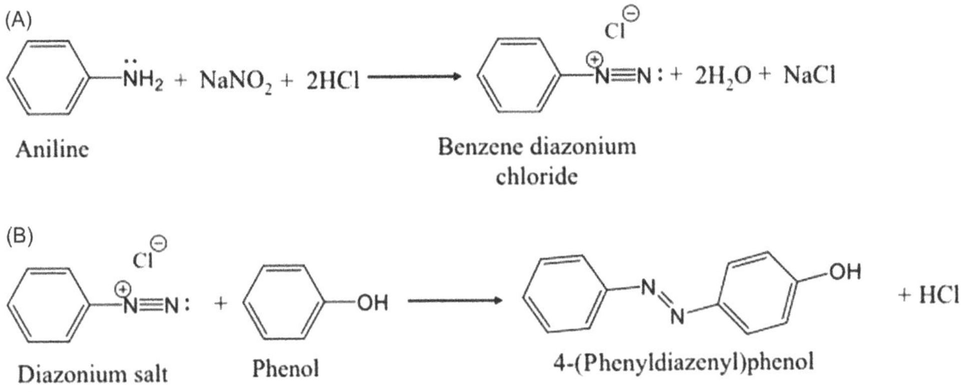

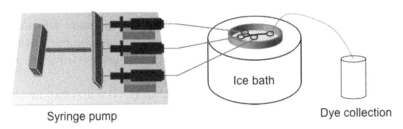

FIGURE 17.1 (A) Scheme for diazotization reaction and (B) scheme for coupling reaction.

FIGURE 17.2 Schematic of a microfluidic synthesis of azo dye setup.

two-step reaction on one reactor chip. The manual handling of the reactive and unstable intermediate is avoided by allowing flow of reactive intermediate. The higher surface-to-volume ratio of microchannel compared to normal batch conditions provides very good temperature control and very rapid mixing in flow method. Furthermore, flow reactions are usually easy to scale up because of their excellent mixing and heat transfer (Fig. 17.2).

Undergraduate students with prior experience in organic laboratories can carry out this experiment. After completing this experiment, students will learn synthesis and characterization of organic dyes using a flow method. This experiment integrates various fields of science such as analytical and organic chemistry and engineering. This project is intended for a ~ 3 hour lab period. Students either can work individually or as a team of $2-3$ students during one laboratory session. The microfluidic device, either glass or PDMS, can either be prepared in lab or purchased commercially (*see* Appendix VI).

17.2 Design of the Microfluidic Device

The microchip used in this experiment is similar to that used in the synthesis of gold nanoparticle except with a third inlet. Aniline and nitrous acid are introduced into the chip from

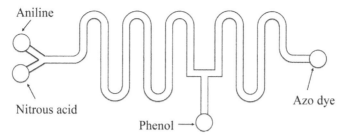

FIGURE 17.3 Channel pattern of the microfluidic device used for dye synthesis.

inlets on one end of the chip, whereas phenol is introduced from the middle of the chip. The dye formed is collected at the opposite end of the chip (Fig. 17.3).

17.3 Chemicals and Supplies

Aniline, nitrous acid, phenol, 10% sulfuric acid, 2 M hydrochloric acid, and 10% sodium hydroxide.

Microfluidic device, syringe pump, syringes, connectors, ice bath/beaker with ice, test tubes, filtration apparatus, melting point apparatus, oven at 105°C, crushed ice, UV–vis spectrophotometer, and IR spectrophotometer.

17.4 Hazards

The reagents in the experiment have certain hazards. The diazonium salt solution is unstable and prone to deteriorate (decompose) when standing at room temperature. The solution should always be kept at below 10°C. Personal protective equipment including safety goggles, gloves, and a lab coat must be worn at all times during the experiment. Long pants should be worn along with close-toed shoes. Be careful when handling the products; they are deeply colored and will stain your skin and clothing for a long period of time. Azo dyes are toxic and may cause genetic mutations. Disposal must follow proper waste-disposal regulations.

17.5 Experimental Procedure

17.5.1 Solution Preparation

Prepare the following solutions in the required volumes:

1. 10% NaOH solution in water
2. Sodium nitrite solution in water
3. Acidic solution of aniline
4. Phenol in 10% NaOH

17.5.2 Dye Synthesis

1. Place the aniline solution in an ice bath.
2. Fill three 1-mL syringes with the individual solutions: aniline, sodium nitrite, and phenol.
3. Connect the microfluidic device with the syringes filled with aniline, sodium nitrite, and phenol solutions.
4. Attach the three syringes to the syringe pump where the plungers are aligned. Remove any excess solution from the ends if syringes need to be adjusted.
5. Select a flow rate. A recommended flow rate is 0.09 mL/min. Inject the solutions from the respective inlets.
6. Acidify the collected product with HCl for the dye to crash out of the solution, if necessary. Filter the solid and rinse twice with water.
7. Collect UV−vis spectra on a small diluted portion of your solution and record the wavelength of the maximum absorbance at different intervals of reaction time.
8. Filter the solid, preferably under reduced pressure, and wash well with water. Remove the filter paper and precipitate, lay it on a watch glass, and put it in an oven at about 105°C.
9. Analyze the solid product by FT-IR.
10. Weigh and calculate the yield of azo dye.
11. Keep your synthetic dye in a properly labeled vial. Dispose the filtrate. Clean your chip by running water through the channel and store it dry.
12. Determine the melting point of the dye and compare with literature value.

17.5.3 Lab Report

Write a lab report following ACS style for journal articles. It should include an abstract, introduction, experimental, results and discussion, and conclusion sections with diagrams, tables, calculations, and images wherever necessary.

Consider the following points when writing the report:

1. Include a short description of flow-based azo-dye synthesis in microchip including its strengths and weakness.
2. Outline the major steps in the experiment.
3. Add representative IR and UV−vis spectra. Do not forget to label the coordinates of all spectra.
4. Include the percent yield of azo dye obtained. Show calculations. Share your personal observations.

17.6 Additional Notes

1. Students should figure out the concentration of aniline, nitrous acid, and phenol. Usually 0.2 M aniline in 1.3 M HCl, and 15% nitrous acid.

2. Let students vary the flow rate of solutions.
3. Combination of other aromatic compounds can be used in this experiment to make different types of dyes. If time permits, synthesis of at least two types of azo dye is suggested. This allows for a comparison between two dyes.

17.7 Assessment Questions

1. Why do you have to keep the temperature at 10°C when preparing the diazonium salt?
2. Write the mechanism for the formation of the diazonium salt. Write a balanced equation for the preparation of the diazonium ion that you made.
3. How would the flow rate affect your reaction?
4. Discuss the advantages of flow chemistry in terms of laboratory and industrial applications.

References

1. Waring DR, Hallas G. In: Waring DR, Hallas G, eds. *The Chemistry and Application of Dyes.* New York: Springer Science & Business Media; 2013.
2. Chudgar RJ, Oakes J, Dyes AZO. *Kirk–Othmer Encyclopedia of Chemical Technology.* Hoboken, NJ: John Wiley & Sons; 2003.
3. Pavia DL, Lampman GM, Kriz GS, Engel RG. *Introduction to Organic Laboratory Techniques: A Small Scale Approach.* 1st ed. Florida: Harcourt Brace; 1998.
4. Feng ZV, Edelman KR, Swanson BP. Student-fabricated microfluidic devices as flow reactors for organic and inorganic synthesis. *J Chem Educ.* 2015.
5. Gung BW, Taylor RT. Parallel combinatorial synthesis of azo dyes: a combinatorial experiment suitable for undergraduate laboratories. *J Chem Educ.* 2004;81(11):1630.
6. Elvira KS, i Solvas XC, Wootton RC. The past, present and potential for microfluidic reactor technology in chemical synthesis. *Nat Chem.* 2013;5:11.

18

Protein Immobilization on a Glass Microfluidic Channel

18.1 Background

Proteins are biological macromolecules that play essential roles in life processes including metabolic process regulation, cellular information exchange, cell-cycle control, molecular transport, and protection from the environment. Many enzymes in our body systems are also protein molecules. The protein immobilization is essential for many advanced chemical and biological applications including identification and use of biomarkers for disease diagnosis, immunoassays, enzymatic reactors, cell sorting, cell migration, DNA sequencing, protein–cell interactions, etc.[1]

Microfluidic bioassay methods require the immobilization of protein molecules on the microchip surface. Protein immobilization involves the attachment of protein molecules onto the microfluidic channel surface. There are various methods available for immobilization, which are primarily divided into physical or chemical methods. Selection of a method depends on the immobilization surface (e.g., glass, PDMS, etc.), sample matrix, buffer type, protein properties, and type of immunoassay and enzyme. The immobilized protein molecules face away from the immobilization surface to mitigate steric hindrance. While choosing the immobilization method, we need to keep in mind that the protein active sites are not sterically blocked by neighboring immobilized proteins. In addition, the active sites for antibody binding or enzymatic conversion should be accessible to reaction partners. After immobilization, protein conformation should be intact so that protein functions are retained for a high-performance reproducible assay.[2]

Major strategies for protein immobilization can be classified into two categories: physical adsorption and chemical covalent immobilization. Physical adsorption methods are based on either passive adsorption onto hydrophobic surfaces, or on electrostatic interactions with charged surfaces. This method is simple and cheap but the interactions that bind the proteins are realtively weak, resulting in gradual desorption. Physical adsorption is therefore less useful for quantitative analytical assays. The second chemical immobilization method involves the formation of covalent bonds between functional groups on the proteins and surfaces. Surfaces with covalently immobilized proteins are very stable, rendering them suitable for quantitative analytical assays. However, the abundance of reactive functional groups on the protein leads to proteins being immobilized in a nonspecific way resulting in random orientation distribution. Therefore covalent attachment may lead to reduced protein activity.

Laboratory Methods in Microfluidics. DOI: http://dx.doi.org/10.1016/B978-0-12-813235-7.00018-0

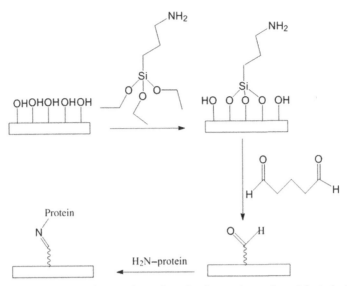

FIGURE 18.1 Schematic of covalent attachment of protein molecules on glass surface. *Adapted with permission from Giri B, Peesara RR, Yanagisawa N, Dutta D. Undergraduate laboratory module for implementing ELISA on the high performance microfluidic platform.* J Chem Educ. *2015;92(4):728–732. Copyright (2015) American Chemical Society.*

Covalent bonds are a frequently used immobilization mechanism in microfluidic assays. Microchannel surface has to be activated for covalent attachment of proteins. The activated surface then reacts with amino acid residues on the protein exterior and forms an irreversible linkage. An enormous variety of covalent conjugation chemistries are available.

In this experiment, we will attach protein molecules on glass microfluidic channel using covalent bonding. The glass surface will be at first reacted with aminosilane molecules like (3-aminopropyl)triethoxysilane (3-APTES). These molecules can bind to hydroxyl functional groups of glass at one end leaving amine groups on the surface at the other end. This surface is then reacted with a cross-linker known as glutaraldehyde. The glutaraldehyde molecule has two reactive aldehyde groups at its two ends and therefore can crosslink two amine functional groups. One aldehyde group of glutaraldehyde group reacts with the amine group on glass surface while the other one is available for reaction with amine groups of protein molecules (*see* Fig. 18.1). The protein immobilized in this experiment is bovin serum albumin (BSA) and it is conjugated with fluorescein isothiocyanate (FITC) dye for visualization.

The goal of this experiment is to provide a protein immobilization protocol suitable to be performed by undergraduate and high-school students in a limited amount of time. This experiment is suitable for a high-school course and can also be integrated into undergraduate chemistry, physics, material sciences, or engineering laboratory courses. After completing this experiment, students will learn immobilization of protein using a microfluidic platform, fluorescence microscopy, surface modification of glass microchannel, and reactivity of organic functional groups in bioconjugation.

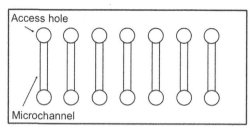

FIGURE 18.2 Design of the microfluidic device.

18.2 Microfluidic Device Design

The glass microfluidic device for this experiment $1'' \times 2''$ with seven microchannels. Each microchannel is 15 mm long, 0.5 mm wide, and 0.03 mm deep. The channels have two access holes (ports) at the end of them. These ports are ~ 1 mm in diameter and are used to introduce and remove fluids (*see* Fig. 18.2).

18.3 Chemicals and Supplies

Deionized (DI) water, sodium hydroxide (NaOH), methanol, 3-APTES, glutaraldehyde, carbonate–bicarbonate buffer, and BSA-FITC.

Microchip, adhesive tape, vacuum pump, oven, pipet and tips, epifluorescence microscope, heater, thermometer, and timer.

18.4 Hazards

Sodium hydroxide is corrosive. 3-APTES may irritate eyes and skin. Eye goggles, gloves, and lab coats are recommended in this experiment. The glass device should be handled carefully.

18.5 Experimental Procedure

18.5.1 Solution Preparation

The following solutions and reagents are required for this experiment:

1. 1 M NaOH
2. 5% (w/v) glutaraldehyde
3. M carbonate/bicarbonate buffer (pH 9.4)
4. 1000 µg/mL stock solution of BSA-FITC and series of five dilutions made in carbonate/bicarbonate buffer.

18.5.2 Surface Modification of Microchannel

Follow the steps given below for the modification of micro channel surface.

1. Introduce a solution of 1 M NaOH into the microchannel ensuring that no air bubbles are trapped within it. Seal the access holes using adhesive tape to prevent any evaporation of this liquid from the channel terminals and incubate for 20 minutes.
2. Remove the NaOH solution from the microchannel using a vacuum pump. Now rinse the channel by flowing DI water through it for 30 seconds. Repeat this washing step with methanol for another 30 seconds. Place the microchip in an oven maintained at 80°C for 10 minutes and later cool it to room temperature.
3. Introduce 3-APTES into channels 1−6 and leave channel 7 with water and seal its access holes using adhesive tape immediately. APTES is hydrolytically unstable and can react with moisture from the atmosphere. Allow APTES to react with the surface of the glass channel for 30 minutes.
4. Remove the excess APTES solution and rinse the channel with methanol for 30 seconds. Dry the channel by placing the microchip in an oven maintained at 80°C for 10 minutes and later cool it to room temperature.
5. Introduce a 5% (w/v in water) glutaraldehyde solution into the conduit, tape its access holes, and incubate for 30 minutes. Remove the excess glutaraldehyde solution and wash the channel with water for 30 seconds.
6. Introduce various concentrations of BSA-FITC solution prepared in carbonate−bicarbonate buffer to 1−5 microchannels.
7. Introduce buffer solution into channel 6 (blank channel).
8. Introduce stock solution of BSA-FITC into the microchannel 7.
9. Seal the access holes with adhesive tape and incubate for 1 hour at room temperature.
10. Wash the channels with the buffer.

18.5.3 Data Collection

1. Place the microchip on the epifluorescence microscope stage
2. Align the relevant microchannel with the 10 × microscope objective to take a fluorescence measurement.
3. Open the mechanical shutter to irradiate the first channel with the excitation beam from the microscope lamp, take a fluorescence image, and then close the shutter immediately.
4. Repeat step 3 for the remaining six channels.

18.5.4 Image Analysis and Quantitation

1. To quantitate the fluorescence images, open them using ImageJ software and measure the average intensity drawing a square box with dimensions about a third to half of the channel width located around the center of the microchannel region (*see* Fig. 18.3).

FIGURE 18.3 Photo of a microchannel. *Adapted with permission from Giri B, Peesara RR, Yanagisawa N, Dutta D. Undergraduate laboratory module for implementing ELISA on the high performance microfluidic platform.* J Chem Educ. *2015;92(4):728−732. Copyright (2015) American Chemical Society.*

2. Subtract the signal of the blank channel from the signal from the protein immobilized channel and the difference is the net signal obtained from protein immobilization.
3. Plot the signal of channels 1−5 on the *y*-axis and corresponding concentration on the *x*-axis. Draw a trend line and obtain regression equation and correlation coefficient.

18.5.5 Lab Report

Write a lab report following ACS style for journal articles. It should include an abstract, introduction, experimental, results and discussion, and conclusion sections with diagrams, tables, calculations, and images wherever necessary.

Consider the following points when writing the report:

1. Include a short description of the principles behind covalent protein immobilization including its advantages over other methods.
2. Outline the major steps and workflow involved in the experiment. Be sure to report the experimental conditions for the assay and the operational parameters for the instrumentations used.
3. Add representative photos of assay channels and calibration curves.

18.6 Additional Notes

1. In order to minimize contamination issues, it is suggested that only one of the channel terminals be used to introduce solutions and the other for draining them out.
2. Care must be taken to ensure that there are no air bubbles trapped in the access hole, which otherwise may obstruct the flow into the channel during the filling process.
3. The vacuum system used to purge reagents from the microchannel may comprise a portable vacuum pump with its vacuum port connected to a pipette tip though plastic tubing.
4. The surface chemistry described in this experimental module is suitable for glass microfluidic devices. But methods for PDMS channels can also be adopted.

18.7 Assessment Question

1. Did you get any signal for channel 7? If yes explain.

References

1. Whitford D. *Proteins: Structure and Function.* 1st ed West Succex, England: Wiley; 2005.

2. Scouten WH, Luong JH, Brown RS. Enzyme or protein immobilization techniques for applications in biosensor design. *Trends Biotechnol.* 1995;13(5):178–185.

3. Giri B, Peesara RR, Yanagisawa N, Dutta D. Undergraduate laboratory module for implementing ELISA on the high performance microfluidic platform. *J Chem Educ.* 2015;92(4):728–732.

Microfluidic Enzyme-Linked Immunosorbent Assay

19.1 Background

Enzyme-linked immunosorbent assays (ELISA)[1] are important bioanalytical techniques used for the quantitative determination of a variety of analytes such as hormones, antibodies, viruses, bacteria, pollutants, etc. This method was developed about 40 years ago and takes advantages of the specific and selective interaction between an antigen and an antibody to detect one or the other in complex sample matrices (e.g., serum) without the need for much sample preparation. The enzyme used in this technique as detection label converts substrate molecules into product molecules. One of the popular formats of ELISA is sandwich ELISA among many other types. In sandwich ELISA, capture antibodies are immobilized on a solid support in the first step. These antibody molecules capture corresponding antigens (analyte) present in the sample. The second antibody conjugated to enzyme label is then used to make the sandwich. The enzyme then converts enzyme substrate molecules into product molecules. The rate of generation of product molecules is then taken as the measure of the concentration of analyte. Because the reaction is enzyme catalyzed, product molecules are generated continuously; it amplifies the assay signal so that low concentrations of analytes can be measured.

Traditionally, ELISA has been performed in 96-well microplates. However, microfluidic devices have emerged as a powerful platform for the implementation of immunoassays in microchannels. Performing ELISA on a microfluidic device[2] offers several advantages over the conventional microplate method. The important benefits of a miniaturized version of the assay method are reduced sample and reagent volume requirement and shorter incubation period with comparable or greater sensitivity in some cases. The reduced volume of ~ 3 orders of magnitude comes from the smaller volume of microchannel compared to the microtiter well. As a result, these miniaturized assays are cheaper than microtiter plate-based assays. Because of the depth of the microchannel, the length scale over which analytes/antibodies need to diffuse to reach the assay surface is over two orders of magnitude shorter than that in a microwell. This surface area-to-volume ratio and diffusion time in combination reduces the incubation period in microfluidic systems by expediting the immunocapture process. In addition, the smaller amounts of sample/reagent used in microfluidic ELISAs reduce the generation of biohazardous waste as well as operator exposure to these materials.

In this experiment,[3] we perform the quantitative determination of anti-bovine serum albumin (BSA) using ELISA in micrometer-sized glass microchannel. Horseradish peroxidase (HRP) will be used as enzyme and Amplex red will be used as enzyme substrate.

Before starting the immunoassay procedure, the microchannel surface has to be modified in such a way that it allows covalent attachment of protein molecules (*see* Chapter 18: Protein Immobilization on a Glass Microfluidic Channel). In case of glass microchannel, the surface is modified with (3-aminopropyl)triethoxysilane (3-APTES) and glutaraldehyde one after the other to have aldehyde groups covalently attach the protein molecules. The micro-fluidic conduits are then reacted with BSA and later incubated with a chosen dilution of an antimouse BSA (analyte) sample. Treating the analysis channels with goat antimouse IgG HRP conjugate completes the ELISA surface (Fig. 19.1).

The enzyme reaction is initiated by introducing Amplex red (enzyme substrate) solution into the fluidic ducts. The HRP enzyme converts colorless and nonfluorescent Amplex red into highly fluorescent resorufin molecules.

We will monitor the enzyme reaction, in this experiment, by taking photos of the assay channel. An epifluorescence microscope connected with a camera is used to take the photos every 5 minutes of the assay reaction. The assay signal is then obtained by processing the photographs using image-processing software. Then, the calibration curve is prepared and concentration of analyte in an unknown sample is estimated. The assay method described in this experiment utilizes saturation kinetics and follows the kinetic ELISA format. Enough enzyme substrate molecules are supplied in saturation kinetics reaction to ensure enzyme reaction is not limited by the substrate.

There are two different ways to monitor the assay signal in ELISA: end-point ELISA and kinetic ELISA. In end-point ELISA, the enzyme reaction is stopped after allowing the assay to run for a given time and the signal is recorded. The reaction is stopped, in many cases, by adding quenching agents (generally acids or bases as enzyme inhibitors). Therefore the end-point ELISA method relies on one data point per sample/assay. Most of the commercial

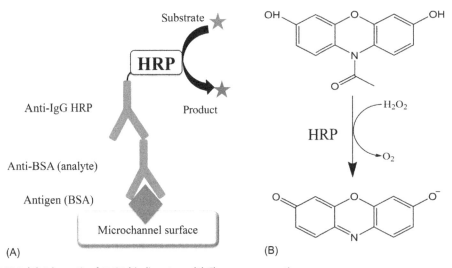

FIGURE 19.1 (A) Schematic of ELISA binding steps. (B) The enzyme reaction.

microtiter plate-format ELISA kits follow end-point ELISA. But this method has several disadvantages as follows[4]:

1. If the assay contents are not mixed well, the quenching may not be complete.
2. Optical properties of product molecule may change when adding acid or base for quenching.
3. It has to rely on a single data point, which can lead to false results.
4. Variations in background signal among the assay chambers lead to inaccurate results.

Most of the disadvantages listed above can be overcome by using kinetic ELISA (k-ELISA). In k-ELISA signal is monitored continuously. A graph is plotted as assay time on the *x*-axis and the signal is plotted on the *y*-axis. The slope of the linear line corresponds to the concentration of the analyte. These linear lines are obtained for different dilutions or concentrations of sample/standard. k-ELISA[5] depends on several data points for each assay, which is more quantitative than just relying on a single point. In k-ELISA it doesn't matter if the background of the assay chamber is different. What matters is the slope of the line. k-ELISA also eliminates the quenching step. By convention, k-ELISA data are expressed as the change in signal over time. The k-ELISA system measures the rate of reaction that leads to development of color. Variability in results is thus reduced because the readings are independent of the time of substrate addition. The slope of the reaction rate is directly proportional to the amount of antibody present in the reaction vessel. There is a general perception among diagnosticians that k-ELISA is superior to single-read ELISA. The advantages of k-ELISA are as follows:

- Less incubation time
- No quenching step
- Wider dynamic range of analyte concentrations
- Allows for checking the linearity of a reaction with time
- No subjective judgment of the straight-line relationship between OD and analyte concentration
- True quantitative results

In this experiment, adapted from reference 3, the enzyme reaction follows a zeroth-order reaction in which the reaction does not depend on the concentration of enzyme substrate. Providing enough substrate concentration ensures a linear increase in assay signal. This experiment is suitable for upper-division undergraduate curriculum of science and engineering. Students will be trained on the kinetic format of the assay, basic image analysis techniques, as well as signal-to-noise ratio and limit of detection calculations that are valuable for characterizing any analytical method. The experiment can be completed in two 3-hour periods (including all incubation and data-collection procedures).

19.2 Microfluidic Device Design

The microfluidic device used in this experiment (Fig. 18.2) is a $1'' \times 2''$ chip containing at least seven microchannels. Each microchannel is 15 mm long, 0.5 mm wide, and 0.03 mm deep. The channels have two access holes (ports) at the end of them. These ports are ~1 mm in diameter and are used to introduce and remove fluids.

19.3 Chemicals and Supplies

Deionized (DI) water, sodium hydroxide (NaOH), methanol, 3-APTES, glutaraldehyde, carbonate-bicarbonate buffer, Tween 20, Amplex Red, hydrogen peroxide, BSA, antimouse BSA, goat antimouse IgG, and HRP.

Glass microchip, adhesive tape, vacuum pump, oven, pipet and tips, epifluorescence microscope, hot-air fan, thermometer, and timer.

19.4 Hazards

This experiment involves the use of several chemicals such as sodium hydroxide, 3-APTES, glutaraldehyde, phosphate and carbonate buffers, Amplex Red, hydrogen peroxide, etc. Eye protection, hand gloves, and laboratory coats are recommended. Contaminated materials should be disposed of appropriately as hazardous chemicals. The edges of the glass microfluidic devices pose a small cutting hazard.

19.5 Experimental Procedure

19.5.1 Solution Preparation

The following solutions and reagents are required in this experiment. It is convenient to have the teaching assistant/instructor prepare and provide some of the reagents.

1. 1 M NaOH, 5% (w/v) glutaraldehyde
2. M carbonate/bicarbonate buffer (pH 9.4)
3. M phosphate buffer (pH 7.4)
4. 1% (w/v) BSA
5. Serial dilution of antimouse BSA standards in phosphate buffer (at least five dilutions)
6. 0.05% (v/v) Tween 20
7. Goat antimouse IgG HRP in phosphate buffer containing Tween 20
8. Enzyme substrate solution containing 10 μM Amplex Red and 5 μM H_2O_2 phosphate buffer

19.5.2 Surface Modification of Microchannel

1. Introduce a solution of 1 M NaOH into the microchannel ensuring that no air bubbles are trapped within it. Seal the access holes using an adhesive tape to prevent any evaporation of this liquid from the channel terminals and incubate for 30 minutes.

2. Remove the NaOH solution from the microchannel using a vacuum pump. Now rinse the channel by flowing DI water through it for 30 seconds. Repeat this washing step with methanol for another 30 seconds. Place the microchip in an oven maintained at 80°C for 10 minutes and later cool it to room temperature.
3. Introduce 3-APTES into the channel and seal its access holes using adhesive tape immediately. APTES is hydrolytically unstable and can react with moisture from the atmosphere. Allow APTES to react with the surface of the glass channel for 30 minutes.
4. Remove the excess APTES solution and rinse the channel with methanol for 30 seconds. Dry the channel by placing the microchip in an oven maintained at 80°C for 10 minutes and later cool it to room temperature.
5. Introduce a 5% (w/v in water) glutaraldehyde solution into the conduit, tape its access holes, and incubate for 30 minutes. Remove the excess glutaraldehyde solution and wash the channel with water for 30 seconds.

19.5.3 ELISA Procedure

1. Introduce a 1% w/v BSA solution prepared in carbonate-bicarbonate buffer, seal the access holes with adhesive tape, and incubate for 1 hour at room temperature. (If the lab session needs a break, it is a good point at which to stop. Keep the BSA-incubated chip in the refrigerator until the next lab session (next day).) Remove the BSA solution and wash the channels with phosphate buffer for 30 seconds.
2. Introduce an appropriate dilution of the antimouse BSA sample prepared in phosphate buffer, tape the access holes, and incubate for 30 minutes. Drain the excess analyte solution and wash the channel with phosphate buffer for 30 seconds.
3. Introduce the goat antimouse IgG HRP solution prepared in phosphate buffer containing Tween 20 into the channel and tape the access holes and incubate for 30 minutes. Remove the excess solution and wash the conduit with phosphate buffer for 30 seconds.

19.5.4 Data Collection

1. Place the microchip on the epifluorescence microscope stage and turn on the heating fan. Using a thermometer, ensure that the air temperature around the microchip reads 37°C.
2. Now introduce the enzyme substrate solution into the channels and seal the access holes with adhesive tape. Remember the enzyme substrate solution must be prepared right before it is used. Start a timer immediately. Align the relevant microchannel with the $10\times$ microscope objective to make a fluorescence measurement.
3. Open the mechanical shutter to irradiate the channel with the excitation beam from the microscope lamp, take a fluorescence image, and then close the shutter immediately. The total beam exposure time should be less than 1 second to prevent photochemical generation of fluorescent species in the enzyme substrate solution.
4. Repeat step 3 every 4 minutes to obtain six images over a 30 minutes enzyme reaction period.

19.5.5 Image Analysis and Quantitation

1. To quantitate the fluorescence images, open them using ImageJ software and measure the average intensity drawing a square box with dimensions about a third to half of the channel width located around the center of the microchannel region (*see* Fig. 18.3).
2. To quantitate the assay, plot the recorded fluorescence intensity in the channel region against the enzyme reaction time and fit the data points to a straight line based on linear regression analysis using Excel.
3. Record the slope of this line and the standard deviation in this slope as calculated by Excel.
4. Now graph the rate of fluorescence generation (slope obtained in step 3) for a sample minus the same quantity for the blank assay against the concentration of the analyte in the sample or reciprocal of the sample dilution factor. This is the response curve for the microfluidic ELISA.

19.5.6 Lab Report

Write a lab report following ACS style for journal articles. It should include an abstract, introduction, experimental, results and discussion, and conclusion sections with diagrams, tables, calculations, and images wherever necessary.

Consider the following points when writing the report:

1. Include a short description of the principles behind ELISA including its advantages in microfluidic device.
2. Outline the major steps and workflow involved in the experiment. Be sure to report the experimental conditions for the assay and the operational parameters for the instrumentations used.
3. Include representative photos of the assay channels and calibration curves.
4. The results of the unknown sample, errors including calculations, and limit of detection for the assay.

19.6 Additional Notes

1. The rate of signal generation in an enzyme assay is highly sensitive to the ambient temperature. Therefore it is better to have constant temperature during the enzyme reaction period in order to get reproducible measurements.
2. In order to minimize contamination issues, it is suggested that only one of the channel terminals be used for introducing solutions and the other for draining them out.
3. Care must be taken to ensure that there are no air bubbles trapped in the access hole, which otherwise may flow into the channel during the filling process.
4. The vacuum system used to purge reagents from the microchannel may comprise a portable vacuum pump with its vacuum port connected to a pipette tip though a plastic tubing.

5. The surface chemistry described in this experimental module is suitable for glass microfluidic devices. These units may also be directly purchased from a commercial vendor for a nominal cost in the absence of an in-house microfabrication facility. It should be possible to adapt the reported ELISA module to polymer microchips that are relatively simple to fabricate.
6. Although the present microfluidic immunoassay describes the quantitative determination of antimouse BSA in a sample, it can be readily applied to detecting other biomolecules of practical interest.

19.7 Assessment Questions

1. What is the role of sodium hydroxide used in the very first step of the ELISA procedure?
2. If the recorded fluorescence in an ELISA is 7.6 after 20 minutes of enzyme reaction and the corresponding value for the background is 2.2, what is the rate of generation of the enzyme reaction product in the assay? Assume that a 1 μM solution of this product species in the assay chamber yields a fluorescence signal of 4.3 and the enzyme reaction follows a zeroth-order kinetics.
3. Using the parameters given in Fig. 19.2 for microtiter plate and microchip, compare the volume and surface area-to-volume ratio of a microtiter well and a microchannel.

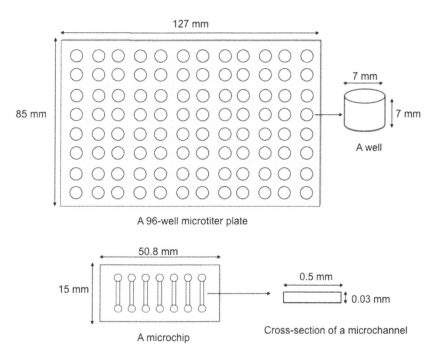

FIGURE 19.2 Schematic of microtiter plate and microchannel showing dimensions. *Adapted with permission from Giri B, Peesara RR, Yanagisawa N, Dutta D. Undergraduate laboratory module for implementing ELISA on the high performance microfluidic platform. J Chem Educ. 2015;92(4):728–732. Copyright (2015) American Chemical Society.*

References

1. Kemeny DM, Challacombe SJ. *ELISA and Other Solid Phase Immunoassays: Theoretical and Practical Aspects*. New York, NY, USA: John Wiley & Sons; 1988.

2. Lee LJ, Yang ST, Lai S, Bai Y, Huang WC, Juang YJ. Microfluidic enzyme-linked immunosorbent assay technology. *Adv Clin Chem*. 2006;42:255–295.

3. Giri B, Peesara RR, Yanagisawa N, Dutta D. Undergraduate laboratory module for implementing ELISA on the high performance microfluidic platform. *J Chem Educ*. 2015;92(4):728–732.

4. Crowther JR. *The ELISA Guidebook*. Totowa, NJ: Humana Press; 2001.

5. Yanagisawa N, Dutta D. Kinetic ELISA in microfluidic channels. *Biosensors*. 2011;1:58–69.

Glossary

ACS	American Chemical Society
AP	4-Aminophenol
APTES	(3-Aminopropyl)triethoxysilane
AU	Artificial urine
BC	Bathocuproine
BCG	Bromocresol green
BOE	Buffered oxide etchant
BSA	Bovine serum albumin
CE	Capillary electrophoresis
COPD	Chronic obstructive pulmonary disease
DI	Deionized
DMG	Dimethylglyoxime
DPC	1,5-Diphenylcarbazide
EF	Electrophoretic
ELISA	Enzyme linked immunosorbent assay
EOF	Electroosmotic flow
EPA	Environmental Protection Agency
FITC	Fluorescein-5-isothiocyanate
GOx	Glucose oxidase
HPLC	High-performance liquid chromatography
HRP	Horseradish peroxidase
HSB	Hue, Saturation, Brightness
IUPAC	International Union for Pure and Applied Chemistry
LOD	Limit of detection
LSPR	Localized surface plasmon resonance
μPAD	Microfluidic paper analytical device
NIH	National Institute of Health
OD	Optical density
PAD	Paper analytical device
PCR	Polymerase chain reaction
PDDA	Poly(diallyldimethylammonium)
PDMS	Poly(dimethylsiloxane)
PEG	Polyethylene glycol
PMT	Photomultiplier tube
RBC	Red blood cells

RGB	Red, Green, Blue
RT	Room temperature
TA	Teaching assistant
TBPB	Tetrabromophenol blue
UV	Ultraviolet

Appendices

Appendix I: Keeping Your Lab Notebook and Writing Lab Report

Every experiment involves two types of recordkeeping: a lab notebook during experimentation and a lab report after completion of the experiment. It is important to write both of these documents in such a way that they are able to deliver the information they are meant to.

I.1 Lab Notebook

The laboratory notebook[1] is a permanent, documented, and primary record of laboratory work and observations. It is not only required for course experiments but also for noncourse experiments (research and service providing experiments). Notebooks are essential tools in many careers, ranging from that of research scientist to that of practicing physician. Even lab-based companies require their employees to keep notebooks with the required information. Any work not recorded in the notebook may not be considered valid. Notebooks are the most important written evidence of completion of a scientific investigation. Moreover, notebooks are the experimental evidence for the conclusion reached, since they are used to document laboratory reports explaining the results. Most importantly, the notebook is a legal document in both academic and industrial settings. In cases of disputes, notebook dates are sometimes used to indicate exactly when an experiment was performed. Ownership of patents can therefore be critically dependent upon keeping a proper notebook. Thus how to keep a laboratory notebook is a major part of most laboratory courses.

A notebook should include everything performed for the experiment in the lab, from procedure to observations. A laboratory notebook should thus be complete, accurate, and detailed enough so that someone who is not familiar with the work can reproduce the same experimental work by following the notes. It can be an excellent resource for reviewing and understanding experiments, writing reports, and deciding the course of future experiments.

I.2 General Guidelines for Lab Notebooks

1. Permanently bound
   a. The notebook should be permanently bound and should not be a looseleaf or in a ring binder.
   b. It may have carbon copies as needed.

    c. The notebook should have consecutively numbered pages, which are considered a permanent part of the notebook. They must not be altered.

2. Permanent ink

    a. Write in your own words using ballpoint pen. Do not use erasable ink and/or pencil.

3. Cover page

    a. Write your name, telephone number, lab section, and email address on the cover page of the notebook.

4. The first pages

    a. Use first couple of pages for a table of contents. Write "Table of Contents" at the top of these pages. The table of contents should include the title of each experiment and the page number on which the experiment begins.

5. A fresh page for each experiment

    a. Start entries at the top of the first page with a heading that includes your name, date, and the title of the experiment. If your notebook has blanks for these items, use the blanks.

    b. Sign the bottom of each page before you leave the lab.

    c. If the experiment is done jointly the notebook must list coworkers and identify who did what.

6. In case of mistakes

    a. Never tear a page out of the lab notebook. If you make a mistake, you should not remove sheets or parts of sheets from your book. In that case, draw a single line through the incorrect entry and then write your correction.

    b. It is often helpful if you very briefly state why you chose to correct the original entry.

    c. If there is a sizable portion of a page that you wish to mark as invalid, draw a diagonal line or large X through that section, and add horizontal lines at the top and bottom to clearly mark the discarded section.

    d. Do not skip pages. If a space is skipped (left blank) in the notebook, it must be marked through with a diagonal line or large X.

    e. Never remove the original pages from the notebook.

7. No scratch paper

    a. It is not appropriate to record data initially on "scratch paper" or any other separate sheet of paper. Any information must be recorded directly in the notebook.

8. Proper labeling

    a. Label all tables, figures, and calculations. It is very unlikely to remember each step or calculation after leaving the lab.

9. No loose paper

    a. There should be no loose scraps of paper in the notebook.

    b. Graphs, charts, spectra, or spreadsheet analyses should be affixed to the pages of the notebook with tape or glue.

10. End of lab day

    a. Initial and date at the end of the notebook page.

I.2.1 Sections in a Standard Notebook

A standard notebook should contain the following sections. Some of these sections must be filled in before the lab starts:

1. *Header*: The header is at the top of the first page of every single experiment. It includes the name of the experiment, date of the experiment, lab section, and your name. You should also identify your partner(s) by first and last names if you are not working alone.
2. *Objective/Purpose*: Write the objective or purpose of the experiment in one sentence prior to the beginning of the lab period. Be as specific as possible.
3. *Reference*: Cite the source upon which the experimental procedure has been followed. In most undergraduate lab courses, it is your lab manual or handout.
4. *Procedure*: This section contains a detailed description of the experimental procedure in chronological order so that someone else could duplicate the experimental work. Also, any code numbers or "IDs" for unknown samples must appear in your lab notebook. Attention to detail in the procedure section will also greatly facilitate the writing of your lab report. Prior to the laboratory experiment, a brief outline of the procedure is required in the notebook as a guideline. But it should NOT be a word-for-word procedure from the lab handout.
5. *Experiment, observations, and data*: This section contains observations and data collected during the experiment. Quantitative observation must include units. Qualitative observations may include color, production of gas during a chemical reaction, formation of precipitate, and many others, i.e., everything you see and hear. Record all data in chronological order. If the experiment involves collection of a large dataset then creating a data table for the data prior to the experiment is suggested. The use of tables will make it much easier for the reader to assess your methods and results. Note any difficulties that you encounter. Drawings of your experimental setup may also be included here. Be sure your data is clearly labeled so that someone else would be able to figure out what it represents.
6. *Results, calculations, and discussions*: It is very important to show each calculation clearly and completely. Pay attention to units, rounding, and significant figures in each number reported. A discussion of the results of an experiment often follows the data and calculation section so that the reader knows the significance of the results. A discussion of the experiment should include qualitative and quantitative comments on the results. Calculations of precision, accuracy, and possible explanations of any obvious errors may be appropriate. If the same calculation is done repeatedly, write one sample calculation in your notebook and report the results of other calculations in a table.
7. *Conclusions*: This section should directly answer the objective of the experiment by writing a brief description of the final result(s).

I.3 Laboratory Reports

A laboratory report is written after the completion of a lab experiment. It is always better to prepare a draft lab report first and then ask your TA or instructor for feedback. The final version of the lab report should incorporate the instructor's suggestions and only then be submitted.

In scientific research, results are reported in the form of scientific papers, which are published in peer-reviewed scientific journals. These papers are not only important in disseminating the results of the research, but are critical for essentially all aspects of career advancement for the scientists involved. Learning to write a proper scientific paper is therefore an important part of the education of all scientists.[2] Considering the fact that you may continue your career in research or service providing labwork in one form or the other, you should write your lab report in a well-defined scientific paper format. The overall format of scientific papers is similar in all journals. In all experiments in this book, the ACS style for journal articles has been recommended. The report should include an abstract, introduction, experimental, results and discussion, and conclusion sections. Include diagrams, tables, calculations, and images wherever necessary.[3]

A laboratory report should contain the following sections:

Title page: This should include the title of the report, your name, your lab partner's name(s), course name and number, lab section, date, and your email.

Abstract: The abstract is the most important element of the lab report. It is like an advertisement for your lab report and readers make initial judgments of your work on the quality of your abstract. Thus it should be a concise and precise summary of the lab report and must be able to stand alone. The typical length of your abstract should be less than 500 words. Therefore in order to present your information to the largest possible audience, an abstract should be clearly written, understandable without having to read the full report, and contain all of the relevant findings from the report. The abstract should start with by telling what the report is all about and the importance of the experiment for a broader audience. It should also state how the work was carried out and important findings. The abstract should end with the larger implications of the experiment.

At the end of the abstract write at least five keywords related to the lab report.

Introduction: This section should include background information of the scientific problem to be addressed and the overall goal of the experiment. You may want to try to answer the following questions: What are you doing? Why and how are you doing it? What was the rationale for the methods you employed? What is the point of the experiment you are about to describe? What strategy are you using to address the experimental question you are asking?

After reading the introduction section, the questions to be addressed in the experiment and why these questions are important should be clear. The introduction section is the place to convince readers why the work is relevant. You may also want to mention limitations and problems you faced during the experiment.

Methods: It should include enough detail of how you performed the experiment so that other students/researchers can reproduce the same experiment. Keep in mind that you are telling a story to people who have not done the experiment. Include a detailed description of materials/equipment and chemicals used including the vendor name and address, your experimental setup, and measurement techniques.

Results and discussion: Results and discussion are usually combined but sometimes they are presented in separate sections. This section should be a description of what you observed and obtained, illustrated with figures and tables. The raw numbers are meaningless unless put them into context. The numerical data should be presented in graphs and tables and put in a context to tell a story. Figures can be very helpful to explain your results. Explain what each figure and table means.

Try to incorporate the goals you were aiming for, what you discovered, which of the results are interesting, and whether your results explain the hypothesis. Also include limitations of the experiment. While putting figures in your report pay attention to the following points:

* Write figure legends clearly so that people simply glancing through the figure can derive useful information from the figures.
* Use one font style and size.
* Avoid shadows, glows, and reflections.
* Do not add a figure with too many datasets. Make less crowded figures and tables.
* Symbols, labels, and scales must be easy to read.
* Include a concise and clear caption for each figure with a figure number.

Conclusion: A clear conclusion helps readers to judge your work easily. Summarize what hypotheses were proved or disproved in your experiment. Indicate uses, extensions, and limitations if appropriate. Suggest future experiments and those that are underway. Indicate explicitly the significance of the work. Do not summarize the report.

Acknowledgments: In this section you should acknowledge those who helped you complete the experiment. The acknowledgment list may include advisors, technical help from your colleague/lab technician/instructor/TA, English editor, suppliers who may have given materials for free, funding agencies (if any), etc.

References: In any scholarly endeavor, it is customary to give credit to the sources of information. The reference section allows you to properly credit the originators of the information you are presenting. Note that unless you invented the method, you should always reference the paper that first described the work.

Appendix: Finally, the report may contain an appendix that contains your raw data and the calculations.

I.2.2 General Criteria for Grading Lab Reports

Your lab report grader may take the following points into consideration while grading the report:

Overall: required sections, clarity in writing, proper use of scientific terms, spelling and grammar, correct calculations, length of report (should not be too long), name of author and lab partner and address on the title page.

Abstract: introduction of overall topic, explanation of hypothesis being tested, description of important methods, conclusion, concise and logical writing.

Introduction: general background, questions to be answered, significance of work, hypothesis, concise and logical writing.

Materials and methods: detail description of experiment, source of the reagents and materials, all methods used.

Results and discussion: rationale of experiment, description of work done, answer and explanation of questions rose in introduction, concise and logical writing, summary of findings, expected results,

Figures: well-designed and understandable, legends, relevancy of figures, title.

Tables: well-designed and understandable, relevancy, description on text, title.

Acknowledgments: clear credits given.

References: proper citing and reference listing, same format to all references.

Appendix: inclusion of raw data and calculations.

Appendix II: Preparation of Selected Reagents and Solutions

II.1 2.0% Ninhydrin Solution[4]

This reagent is used for the determination of amino acids.

Materials and chemicals: Ninhydrin, $SnCl_2 \cdot 2H_2O$, DI water, 100 mL beaker, filter paper, funnel, and watch glass.

Procedure:

1. Mix 1.0 g of ninhydrin and 40 mg of $SnCl_2 \cdot 2H_2O$ in 25 mL of water in a beaker or flask.
2. Cover the mixture and let it stand overnight.
3. Filter the mixture by gravity filtration.
4. Transfer the filtrate to a volumetric flask and dilute to 50 mL with water.

II.2 Buffer Solutions

A buffer is a mixture of a weak acid and its conjugate base (or salt). A buffer solution does not allow a large change in the pH with the addition of small amounts of acids (H^+) or bases (OH^-). Buffer solutions are used in many chemistry, biochemistry, and environmental chemistry experiments to maintain constant pH of the reaction mixture or reagent. The buffer solutions can be purchased from commercial vendors or can be prepared in the lab.

A weak acid (HA) dissociates to give hydrogen ions (H^+) and conjugate base (A^-) as follows:

$$HA \leftrightarrow H^+ + A^-$$

A buffer system for above acid is described by the Henderson–Hasselbalch equation:

$$pH = pK_a + \log_{10}\left(\frac{[A^-]}{[HA]}\right) \tag{A.1}$$

where K_a is the dissociation constant of the weak acid. This equation is used to calculate the pH of a solution and buffer composition of the solution. A buffer has the highest buffering capacity at its pK_a value. The buffering capacity, also known as the efficiency of a buffer in resisting changes in pH, of a buffer is defined as the amount of acid or base needed to change the pH of one liter of buffer solution by one unit. Therefore buffer solutions are made from the weak acid and its salt whose pK$_a$ value is close to the desired pH of the buffer.

Generally 10–100 mM buffer solutions are used. You can make a concentrated buffer solution and then dilute it according to the need of the given experiment. The Henderson–Hasselbalch equation is used to calculate the amount of weak acid/base and its salt to be mixed in a given volume of water.

2a. Preparation of phosphate buffer: Phosphate buffer is one of several very common buffers. It consists of a mixture of monobasic dihydrogen phosphate and dibasic monohydrogen phosphate. By varying the amount of each salt, a range of buffers can be prepared between pH 5.8 and pH 8.0. Phosphates have a very high buffering capacity and are highly soluble in water.

Materials and chemicals: Sodium phosphate dibasic (Na$_2$HPO$_4$·12H$_2$O), Sodium phosphate monobasic (NaH$_2$PO$_4$·2H$_2$O), beakers, graduated cylinders, hot plate, scale, spatulas, stir plate, weigh boats, and pH meter.

Procedure:

1. Prepare sodium phosphate dibasic stock (0.3 M) by dissolving 53.7 g of sodium phosphate dibasic in a final volume of 500 mL of H$_2$O. You may need to warm up the solution for complete dissolution.
2. Prepare sodium phosphate monobasic stock (0.3 M) by dissolving 23.4 g of anhydrous sodium phosphate monobasic in a final volume of 500 mL of H$_2$O.
3. Bring the 0.3 M sodium phosphate dibasic solution from Step 1 to pH 8.0 (for amino acid determination) by adding needed volume of the 0.3 M sodium phosphate monobasic solution from Step 2 while measuring the pH using a pH meter. The resulting solution is 0.3 M phosphate buffer.

If you do not have a pH meter, mix two phosphates as given in Table A.1 to get the desired pH at 25°C. The ratio given in the table can be calculated using the Henderson–Hasselbalch equation.

You may want to dilute the 1 M buffer solution as required to 0.1 M or 0.3 M or other. The carbonate/bicarbonate buffer can be prepared by using a similar method.

II.3 Griess Reagent

Griess reagent is used in the determination of nitrite ions. This indicator should be prepared immediately before class and stored in an amber bottle to protect it from light at 2–6°C.

Materials and chemicals: *p*-amino benzene sulfonamide, N-(1-naphthyl)ethylenediamine, citric acid, methanol, water, and 100 mL volumetric flask.

Procedure[5]:

1. Dissolve 0.86 g *p*-amino benzene sulfonamide, 0.19 g N-(1-naphthyl)ethylenediamine, and 6.34 g citric acid in 20% (*V:V*) water in methanol solution in a 100 mL volumetric flask.
2. Dilute the mixture up to the mark.

Table A.1 The Volume of Dibasic and Monobasic Phosphate Required Making 1 M Buffer Solution of Given pH

pH	Volume of 1 M Na_2HPO_4 Required (mL)	Volume of NaH_2PO_4 Required (mL)
5.8	8.5	91.5
6.0	13.2	86.5
6.2	19.2	80.8
6.4	27.8	72.2
6.6	38.1	61.9
6.8	49.7	50.3
7.0	61.5	38.5
7.2	71.7	28.3
7.4	80.2	19.8
7.6	86.6	13.4
7.8	90.8	9.2
8.0	94.0	6.0

Appendix III: Use and Calibration of Micropipettes

Micropipettes with disposable tips are frequently used to accurately measure small volumes of liquids. Such micropipettes are commercially available in a variety of volumes typically ranging from 0.5 to 1000 μL. There are two types of micropipettes: fixed volume and variable volume. Most of the experiments in this book require the use of micropipettes. The ability to accurately and reproducibly measure and transfer small volumes of liquids is critical to obtain useful results. Therefore we will learn the basics of a micropipette including its calibration here.

III.1 Anatomy of a Micropipette

Micropipette parts may vary to some extent in different types of micropipettes but basic parts are the same. The parts of a typical micropipette are plunger button, tip ejector, pipette tip, tip eject shaft, volume readout, and volume adjustment (Fig. A.1).

III.1.1 Guidelines to Proper Use of a Micropipette

Check your pipette at the beginning of your workday for dust and dirt on the outside. If needed, wipe with 70% ethanol. There are two "stops" when you push the plunger. When you push down gently on the plunger of the micropipette, you will feel a "stop" where the resistance increases. If you push a little harder, the plunger will move even further to a second stop. The first stop is used to suck up the correct volume. The second stop is used to completely expel the liquid you are measuring. Take the following steps to use a micropipette (see Fig. A.2):

1. Put a tip onto the pipette.
2. Set the required volume to be dispensed using a volume adjuster on the pipette.

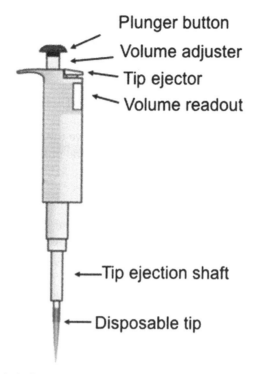

FIGURE A.1 Schematic of a typical micropipette showing various parts.

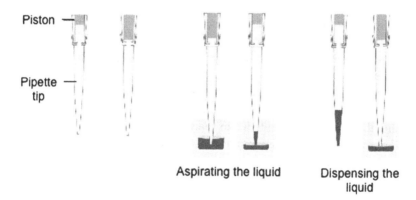

Aspirating the liquid   Dispensing the
                                    liquid

FIGURE A.2 Drawing showing steps in the use of a micropipette. *Adapted from 6.*

3. Prerinse the tip of the pipette by aspiring and dispensing the setting volume three times and removing any remaining liquid.
4. Press the push button to the first positive stop to aspire the liquid.
5. Hold the pipette vertically and then immerse the tip so that the tip is 1−4 mm in the liquid.

6. Release the push button slowly and smoothly to its top position to aspirate the liquid.
7. Wait 1 second and withdraw the tip from the liquid.
8. Wipe any droplets away from the outside of the tip using a Kim wipe.
9. Place the end of the tip against the inside wall of the vessel where you want the liquid to go at an angle of $10-40°$.
10. Press the push button smoothly to the first stop. Wait 1 second, change to the new site, and press the push button on the second stop.
11. Keeping the push button pressed to the end, remove the pipette by drawing the tip along the inside surface of the vessel, and release the push button.
12. Finally eject the pipette tip.

III.1.2 Working Principle of Micropipettes

There are two types of pipettes: air-displacement and positive-displacement pipettes. Air-displacement pipettes are generally used with aqueous solutions. Positive-displacement pipettes are used for highly viscous and volatile liquids. Both types of pipettes have a plunger (piston) that moves in a cylinder or capillary.

In air-displacement pipettes, there remains a layer of air that separates the piston from the liquid being aspired. In positive-displacement pipetting, the piston is in direct contact with the liquid.

III.1.3 Tips for Micropipette Use

Pipettes are precision instruments. They give more consistent results when operated with care. The following tips[7] may help to properly use pipettes for accurate results:

- Never draw liquid into the barrel of the micropipette itself. Always use a pipette tip and right type of tip for the given pipette.
- Do not set the micropipette volume below or above the range given for the pipette under any circumstances.
- Prerinsing the tip with the liquid to be pipetted improves accuracy. Prewetting increases humidity in the tip, thus reducing the amount of and variation in sample evaporation.
- Avoid turning the pipette on its side when there is liquid in the tip. Liquid might get into the interior of the pipette and contaminate the pipette.
- Always store pipettes in an upright position when not in use.
- Allow liquids and equipment to equilibrate to ambient temperature. The volume delivered varies with air pressure, relative humidity, and vapor pressure of the sample, all of which are temperature dependent.
- Wipe the tip carefully and only if there is liquid on the outside. Otherwise, sample liquid may be wicked from the tip.
- Pause with the tip in the liquid for one to two seconds after aspirating the sample. This is important because the liquid in the tip bounces slightly when the plunger stops.
- Do not touch the tip to the sides of the container. Surface tension causes the sample to vary if the exit angle varies, particularly for small volumes.

- Set the pipette down between deliveries. Body heat transferred to equipment during handling disrupts temperature equilibrium.
- Immerse the tip 1–4 mm below the meniscus and well clear of the container walls and bottom during sample aspiration; otherwise volume is affected.
- Securely attach a high-quality tip designed for use with the pipette and appropriate for the size of the container.
- Depress and release the plunger smoothly.

III.1.4 Calibration of Micropipette

Micropipettes are usually quite accurate. However, they may eventually develop problems with use. It is therefore important to calibrate the micropipettes every now and then. Generally pipettes are calibrated every 3 months. Calibration of micropipettes can be an independent experiment for the course and students can be asked to perform this experiment at the beginning of the lab course.

Calibration of pipettes means determining the difference between the selected volume and the dispensed volume. The pipettes are calibrated by gravimetric method in which the amount of water dispensed at a set volume is weighted on the analytical balance. The volume of the water is then back calculated using the temperature at the time of measurement and density of water. Under a constant temperature and atmospheric pressure, the density of water is constant. These factors are usually combined to give the Z factor used in calculation of volume of water. The calculated volume of water is then compared with the theoretical volume to determine the accuracy and precision of the pipette.

A number of factors affect the performance of pipettes such as ambient temperature, relative humidity, density of liquid to be handled, atmospheric pressure, etc. Error in pipettes is quantitated by statistical terms such as accuracy and precision (see Appendix IV for detail description). In brief, accuracy is a measure of how close a measured volume is to the set volume. It is related to the percent error between the average volume of solution measured experimentally and the volume that was expected (the accepted value). The farther from the correct volume, the lower the accuracy of the pipettes and/or the technique. Precision is the measure of reproducibility of a measurement. Precision is related to the standard deviation of a series of measurements of the same thing. The smaller the standard deviation, the more precise the micropipette is. A pipette is always adjusted for delivery of the selected volume. If the calculated results are within the selected limits, the adjustment of the pipette is correct.

A pipette is calibrated for at least two volumes: the maximum volume (nominal volume) and the minimum volume or 10% of the maximum volume, whichever is higher. For example, a 0.5–10 μL pipette is tested at 10 μL and 1 μL. Prior to beginning calibration, be sure that all components, pipettes, tips, balance, and test liquid are temperature stabilized.

1. *Materials and equipment for calibration*: Pipette and tips, 50 mL beaker, plastic medicine cup, distilled water, temperature meter ($\pm 0.1^\circ$C), analytical balance (± 1.0 mg), and atmospheric pressure meter.

2. *Calibration procedure*:
   a. Determine the water temperature and record it.
   b. Place a beaker containing distilled water into the analytical balance and close the door of the balance. Wait for sometime so that indoor environment of the balance gets equilibrium with water vapor.
   c. Place a plastic medicine cup on the pan and adjust the weight to zero.
   d. Put a tip onto the pipette and set the volume to be tested.
   e. Deliver the measured water onto the cup. Close the door of the balance and record the value on the balance display after it has stabilized.
   f. Repeat this procedure for ten times.
   g. Eject the tip.
3. *Calculation in calibration*:
   a. Convert the weight unit of the measured value (w) into the volume unit of the measured value (V) using:

$$V = (w + e) \times z \tag{A.2}$$

   where e = evaporation loss (mg) and z = conversion factor (mg/μL) (see Table A.2).
   Evaporation loss can be significant with low volumes. To determine mass loss, dispense water into the weighing vessel, note the reading, and begin timing with a stopwatch. Determine the weight loss for 30 seconds assuming that pipetting time is 10 seconds. If an evaporation trap or lid on the vessel is used, an evaporation correction is unnecessary. The conversion factor z is for calculating the density of water suspended in air at a test temperature and pressure.
   b. Calculate the mean volume $(\overline{V})$, accuracy, standard deviation (SD), and imprecision (CV).
   Accuracy (systematic error, A) is the difference between the dispensed volume and the selected volume of a pipette and is given by:

$$A = \overline{V} - V_\circ \tag{A.3}$$

   where $V_\circ$ is the set volume.
   Accuracy can be expressed as a relative value as given in:

$$A\% = \frac{A}{V_\circ} \times 100\% \tag{A.4}$$

   Precision (random error) refers to the repeatability of the pipetting. It is expressed as SD or CV (Eq. (A.5)). See Appendix IV for calculation of SD.

$$CV = \frac{SD}{\overline{V}} \times 100\% \tag{A.5}$$

   c. Accuracy value must be 99–101% and CV value must be less than 1%.
   d. Once you determine the inaccuracy and imprecision of your pipette you may need to make adjustments to the unit. Follow the manufacturer's instructions for adjustment.

Table A.2 *Z* Correction Factors for Distilled Water as a Function of Temperature and Atmospheric Pressure for Distilled Water[6]

B (kPa)	80	85	90	95	100	101.3	105
t (°C)				Z (µL/mg)			
15.0	1.0017	1.0018	1.0019	1.0019	1.0020	1.0020	1.0020
15.5	1.0018	1.0019	1.0019	1.0020	1.0020	1.0021	1.0021
16.0	1.0019	1.0020	1.0020	1.0021	1.0021	1.0021	1.0022
16.5	1.0020	1.0020	1.0021	1.0021	1.0022	1.0022	1.0022
17.0	1.0021	1.0021	1.0022	1.0022	1.0023	1.0023	1.0023
17.5	1.0022	1.0022	1.0023	1.0023	1.0024	1.0024	1.0024
18.0	1.0022	1.0023	1.0023	1.0024	1.0025	1.0025	1.0025
18.5	1.0023	1.0024	1.0024	1.0025	1.0025	1.0026	1.0026
19.0	1.0024	1.0025	1.0025	1.0026	1.0026	1.0027	1.0027
19.5	1.0025	1.0026	1.0026	1.0027	1.0027	1.0028	1.0028
20.0	1.0026	1.0027	1.0027	1.0028	1.0028	1.0029	1.0029
20.5	1.0027	1.0028	1.0028	1.0029	1.0029	1.0030	1.0030
21.0	1.0028	1.0029	1.0029	1.0030	1.0031	1.0031	1.0031
21.5	1.0030	1.0030	1.0031	1.0031	1.0032	1.0032	1.0032
22.0	1.0031	1.0031	1.0032	1.0032	1.0033	1.0033	1.0033
22.5	1.0032	1.0032	1.0033	1.0033	1.0034	1.0034	1.0034
23.0	1.0033	1.0033	1.0034	1.0034	1.0035	1.0035	1.0035
23.5	1.0034	1.0035	1.0035	1.0036	1.0036	1.0036	1.0037
24.0	1.0035	1.0036	1.0036	1.0037	1.0037	1.0038	1.0038
24.5	1.0037	1.0037	1.0038	1.0038	1.0039	1.0039	1.0039
25.0	1.0038	1.0038	1.0039	1.0039	1.0040	1.0040	1.0040
25.5	1.0039	1.0040	1.0040	1.0041	1.0041	1.0041	1.0042
26.0	1.0040	1.0041	1.0041	1.0042	1.0042	1.0043	1.0043
26.5	1.0042	1.0042	1.0043	1.0043	1.0044	1.0044	1.0044
27.0	1.0043	1.0044	1.0044	1.0045	1.0045	1.0045	1.0046
27.5	1.0045	1.0045	1.0046	1.0046	1.0047	1.0047	1.0047
28.0	1.0046	1.0046	1.0047	1.0047	1.0048	1.0048	1.0048
28.5	1.0047	1.0048	1.0048	1.0049	1.0049	1.0050	1.0050
29.0	1.0049	1.0049	1.0050	1.0050	1.0051	1.0051	1.0051
29.5	1.0050	1.0051	1.0051	1.0052	1.0052	1.0052	1.0053
30.0	1.0052	1.0052	1.0053	1.0053	1.0054	1.0054	1.0054

Assessment questions:

1. Did the pipet deliver the volume of water expected?
2. Was the pipet accurate?
3. Was the pipet precise?
4. On a computer spreadsheet, plot a graph by putting "volume dispensed" on y-axis and "set volume." Fit a trend line for this data using linear regression and find out the equation of best fit and the correlation coefficient.

Appendix IV: Statistical Treatment of Data

Every measurement made in a laboratory may contain error(s). Two types of errors will occur: systematic and random. Systematic or determinate errors are reproducible errors that can be corrected. Examples are errors due to a miscalibrated piece of glassware or a balance. Random or indeterminate errors are due to limitations of measurement that are beyond the experimenter's control. These errors cannot be eliminated and lead to both positive and negative fluctuations in successive measurements. Examples are a difference in readings by different observers, or the fluctuations in equipment due to electrical noise.

Accuracy and precision are two widely used terms while judging the quality of data obtained. Accuracy describes how close a result is to the true value. It is an agreement between a measured value and the accepted true value. Data that is highly accurate suggests that there is little systemic error. Precision describes how close the results from different trials are to each other. It is the repeatability of your measurements. Data of high precision indicates small random errors and leads experimenters to have confidence in their results. A well-designed experiment (and a well-trained experimenter) should yield data that is both accurate and precise (Fig. A.3).

In an effort to describe and quantify the random errors an average, a standard deviation, a 90% confidence limit, and a relative deviation are generally reported.

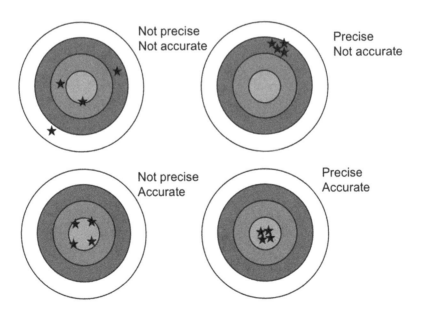

FIGURE A.3 Concept of accuracy and precision. The star indicates the values you obtain from the experiment. The central circle represents the target result you may want to obtain.

IV.1 Average and Standard Deviation

The average or mean, $\bar{x}$, is defined as in Eq. A.6:

$$\bar{x} = \frac{1}{N}\sum x_i \qquad (A.6)$$

where each x_i is one measurement and N is the number of trials or samples.

The standard deviation, SD or s, measures how close values are clustered about the mean. This term quantifies the amount of variation in a set of data values. The standard deviation for small samples is given by:

$$SD = \sqrt{\frac{\sum (x_i - \bar{x})^2}{N-1}} \qquad (A.7)$$

The smaller the value of SD the more closely packed the data is about the mean, and the measurements are considered to be more precise.

IV.2 Calibration Curve and Regression Line

A calibration curve is a plot of measured response (or signal) to a known amount of analyte, the standards. In this case, signal is plotted on y-axis as a response variable and concentration of analyte as explanatory variable is plotted on x-axis. When data is plotted as a scatterplot, drawing a line through the scatterplot can summarize the overall pattern of the response. A least-square regression line (simply a regression line or trend line) is a straight line that describes how a response variable y changes as an explanatory variable x changes. This line is considered as a mathematical model that can be used to predict the value of y for a given x.

The regression equation obtained is as given by:

$$y = mx + b \qquad (A.8)$$

When the regression line passes through all data points, it is a perfect relation between x and y but that rarely happens. Using this equation we can predict the values in y for a given value of x and vice versa. Here, m is called the slope of the regression line and gives the change in y per unit change in x. And, b is known as the y intercept, which is the value of y when x is zero. In a perfect curve where the signal of blank is subtracted from the signal of the sample, the intercept value (i.e., b) is zero. That means the line passes through the origin.

The linearity of the calibration curve is measured by a parameter, R^2, known as the square of the correlation coefficient. A value close to 0 R^2 tells us that the regression line is not a good model for the given data and the y values and x values are not correlated at all. However, a value close to 1 R^2 tells us that the line fits really well with the given data.

Therefore you should always include R^2 as a measure of how successful the regression was in explaining the response when you report a regression line.

The above regression equation is then used to back calculate the value of an unknown in a sample. Once you know the y value of your sample measurement then Eq. (A.8) can be modified as Eq. (A.9) to calculate your unknown:

$$x = \frac{y - b}{m} \tag{A.9}$$

IV.3 Limit of Detection

The detection limit or lower limit of detection (or just LOD) is the lowest quantity of an analyte that can be distinguished from the absence of that substance (a blank value) within a stated confidence limit.

The International Union for Pure and Applied Chemistry (IUPAC) defines[8] LOD as follows:

> The limit of detection, expressed as the concentration, C_L, or the quantity, q_L, is derived from the smallest measure, χ_L, that can be detected with reasonable certainty for a given analytical procedure. The value of χ_L is given by:

$$\chi_L = \overline{\chi}_{bi} + \kappa S_{bi} \tag{A.10}$$

where $\overline{\chi}_b$ is the mean of the blank measures, S_b is the standard deviation of the blank measures, and κ is a numerical factor chosen according to the confidence level desired. The value of κ generally used is 3 for a $\sim 99\%$ confidence.

The χ_L limit of the response value is then converted into a limit of concentration value using a regression equation of the calibration curve.

Assessment questions:

1. Total amino acid content in a tea sample from Nepal was found to be 29.2 mg/g, 31.5 mg/g, 30.0 mg/g, 29.8 mg/g, and 32.5 mg/g. Report the amount of amino acid on average and its standard deviation.

Appendix V: Installation and Use of ImageJ Software

ImageJ is a freely available image analysis computer program developed by the National Institute of Health (NIH).[9] It runs on all platforms that support Java. You can download this software compatible for Mac OS X or Linux or Windows from http://rsbweb.nih.gov/ij/ and then install the program. ImageJ can display, edit, analyze, process, save, and print 8-bit, 16-bit, and 32-bit images. It can read images in many formats such as TIFF, GIF, JPEG, BMP, DICOM, TIFS, and raw.

In our experiments the images are taken as color image. The RGB model is used to describe the color in these images. RGB (Red, Green, and Blue) is the most commonly used color space. Another color space we will need is HSB (Hue, Saturation, and Brightness)

V.1 How to Use ImageJ?

1. Download and install the ImageJ program.
2. Locate the ImageJ program file and start the program. You should be greeted with the floating window as shown in Fig. A.4.
3. Open a file:
   File→Open (or press Ctrl + O). Use the dialog box to locate your file.
4. Quantitative measurements of "color":
   - Convert to grayscale:
     Image → Type → 32-bit. The result will be a 32-bit grayscale version of the image. Since the grayscale values one can get from the converted image is in opposite way: darker spots give lower values than brighter spots. The calibration curve obtained by these values will have a negative slope. Therefore it is always good to invert the image.
   - Invert the colors:
     Edit → Invert (or press Ctrl + Shift + I). We are now ready to make measurements.
   - Make the measurement:

   Select the region of the image to analyze using one of the selection tools in the *ImageJ* menu.

   Analyze → Measure (or press Crtl + M).

A results window will appear with the area of the selection and the mean, max, and min grayscale values of the selection.

Move the same selector to another measuring section so that you get the same area of measurement (Figs. A.4 and A.5).

Applying color threshold: A color threshold is applied to see only the color of interest. For each assay color, the hue is adjusted to a specific range, within which colors not attributed to the desired assay color are filtered out.

Image → Adjust → Color Threshold. In this window HSB is selected (at bottom), which allows adjustment of hue, saturation, and brightness.

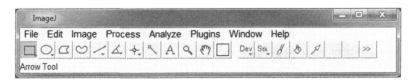

FIGURE A.4 Screenshot of ImageJ window.

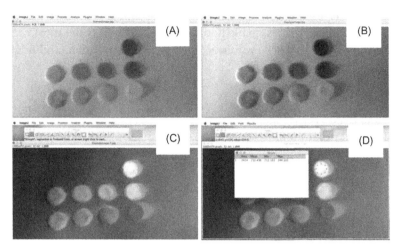

FIGURE A.5 Screenshots of (A) an image opened with ImageJ, (B) same image converted into grayscale, (C) inverted grayscale image, and (D) ImageJ screenshot showing measured values.

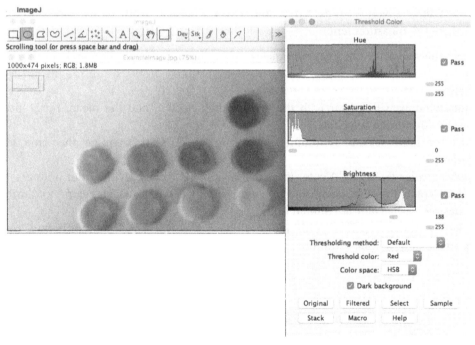

FIGURE A.6 ImageJ screenshot showing hue adjustment.

- Hue adjustment:
  Move the sliders directly below the "Hue" spectrum until only the color of interest is visible. The hue threshold range may be different for different assays.
- Convert to grayscale:
  Image → Type → 8-bit
- Invert the colors:
  Edit→ Invert
  Measure the gray value by selecting the detection zone in the assay image (Fig. A.6).

Appendix VI: Where to Buy: Vendors and Service Provider Information

Item	Model/Description	Manufacturer/Vender
A. Equipment/Instrument		
Wax Printer	ColorQube 8570, Phaser8860, ColorQube 8870	Xerox corporation
Desktop Flatbed Scanner	DocuMate 3220	Xerox corporation
	CanoScan LiDE 210	Canon Inc., Tokyo, Japan
	Perfection 4990, Perfection 1640	EPSON
Sputtering System	Sputtering DC model BA 510	Balzers, Germany
	NTE-3000/4000	Nano-Master, TX, USA
pH/Ion Analyzer	Model 350	Corning, USA
	Orion Star™ A211	Thermo Scientific
	sympHonyTM B10P	VWR International, USA
Potentiostat	PGSTAT-30 Model	Autolab Eco Chemie, The Netherlands
	ED40 Electrochemical Detector	Dionex, USA
Water Purification	MilliQ Millipore system	Millipore, USA
	Barnstead MicroPure	Thermo Scientific
UV Lamp	High-Intensity UV Inspection Lamps, UVP®	VWR International, USA
	UVP™ Blak-Ray™ B-100A UV Lamps	Fisher Scientific
Oven	HERAthem Advanced Protocol Ovens	Thermo Scientific
	VWR® Forced Air Ovens	VWR International, USA
XY Profiler	Zeta-20 Benchtop Profiler	Nano Science Instruments, USA
	DektakXT® Stylus Profiler	Bruker
	Alpha-Step D-600	KLA Tencor
Hot Plate with Stirrer	Cimarec digital hotplates	Thermo Scientific
	Tecnal TE-038	Tecnal, Piracicaba, Brazil
Sand Blaster	Master Problast 3-80060, Problast 3-80070	Vaniman Manufacturing company
	ALC® Polymer Benchtop Abrasive Blaster	Northern tools

(Continued)

(Continued)

Item	Model/Description	Manufacturer/Vender
Vortex	BR-2000 Vortexer	Bio-Rad
	Fisher Scientific™ Analog Vortex Mixer	Fisher Scientific
Syringe Pump	Harvard PHD ultra 22/2000 syringe pumps	Harvard Apparatus
	KDS 101 Legacy syringe pump	KD Scientific Inc.
	NE-1002X	New Era Pump System, NY
	Fusion 100/200	Chemyx Inc., TX
	MFCS™-EZ	Fluigent, Inc., MA
	AF1 Duel—Vacuum & pressure controller	ElveFlow, France
Inverted Epifluorescence Microscope	TE 2000 S	Nikon, Melville, NY
Preamplifier	SR-560	Stanford Research Systems, CA
High-Voltage Power Supplies	PHV400, PS 310 and PS 350	Stanford Research Systems, CA
	HVS448	LabSmith, CA
Photo Multiplier Tube (PMT)	H9 306-02	Hamamatsu, Bridgewater, NJ
Laser Light Sources	–	Melles Griot
CCD Camera	–	Roper scientific
B. Selected Reagents and Samples		
Certified Reference Material for Metal	Trace metals-fly ash 2 #CRM012	Sigma Aldrich
Citrated Rabbit Whole Blood	RB*050	HemoStat Laboratories, CA
	Rabbit Whole Blood – Sodium Citrate (New Zealand White)	BioreclamationIVT
	030-ABSC-PMG	BioChemMed Services, USA
Human Serum Samples	Humatrol N and Humatrol P	HUMAN, Wiesbaden, Germany
	Hypo-Opticlear Human Sera SAE0012	Sigma Aldrich
	Human serum P30-2401	PAN-Biotech
Antimouse BSA	ab19194	Abcam
	5F9 IgG1	Rockland Immunochemicals Inc.
Goat Antimouse IgG-HRP	A9044	Sigma Aldrich
Bovine Serum Albumin (BSA)	A0281, A8531	Sigma Aldrich
Protein	P0914	Sigma Aldrich
Glucose	G8270	Sigma Aldrich
Glucose Oxidase	G2133	Sigma Aldrich
Artificial Urine	Clear Choice® Sub-Solution	Ventures LLC, USA
Triacetin Oil (glyceryl triacetate)	W200700	Sigma Aldrich
Light Mineral Oil	330779	Sigma Aldrich
Albumin—Fluorescein Isothiocyanate Conjugate	A9771	Sigma Aldrich
Jell-O Jelly Powder	–	Wal-Mart, Amazon.com
Gelatin	–	Wal-Mart, Amazon.com
Citrate Buffer	C2488	Sigma Aldrich

(Continued)

(Continued)

Item	Model/Description	Manufacturer/Vender
C. Supplies		
PDMS Chemicals	Sylgard 184 Silicone Elastomer Kit	Amazon.com
	SYLGARD® 184	Sigma Aldrich
	RTV 615	R. S. Hughes Company Inc. CA
Biopsy Punch	Punch, Biopsy (2mm)	Premier Medical
	Robbins Instruments Disposable Biopsy Punches	Amazon.com
Photo Developing Solution	Microposit developer MF-319	Rohm and Haas
	SD-1	Tokuyama, Japan
	Positive developer 418	MG chemicals
Chromium Etchant Solution	Chromium etchant 1020	Transense Inc., USA
	Chromium etchant 651826	Sigma Aldrich
	Chrome etch no 1	MicroChemicals, Germany
Buffer Oxide Etchant (BOE)	BD Etchant	Transense Inc., USA
	BOE 7:1	TECHNIC FRANCE
	CMOS 6:1	AVANTOR PERF MAT-ELECTR PROD
Glass Substrate Coated With Chromium and Photoresist	–	Telic Company, CA, USA
Cover Plate Glass	–	Telic Company, CA, USA
		Chemglass Life Sciences, NJ, USA
		Nanofilm, CA, USA
Glass Cutter	Zoro #G8665544	www.zoro.com
	#8501	The Home Depot, www.amazon.com
Plastic Tweezer	Nalgene™ Polypropylene Scissor-Type Forceps	Thermo Fisher
Magnetic Stirrer Bar	A467	www.amazon.com
Pipettes	0.5–10 μL: # AQ61880	BIORAD
	2–20 μL: # AQ02455	
	20–200 μL: # OH19528	Fisher brand
	100 μL–1000 μL: #, OH32899	
Syringes	Gastight, 1005 C, 5.0 mL	Hamilton glass syringes
UV Curable Glue	NOA 68 T, NOA 68TH	Norland Products Inc.
	PARLITE® UV ADHESIVES	Parson Adhesives Inc., USA
	UV10TKMed	Master Bond Adhesives, USA
Band-pass Optical Filters	–	Semrock Inc., USA
Whatman Filter Paper	Number 1, P81	GE Healthcare Life Sciences
Blood-Separation Filter	LF1, MF1	GE Healthcare Life Sciences

(Continued)

(Continued)

D. Service Providers

Item	Manufacturer/Vendor
Glass-Based Chips	The Dolomite Centre Ltd., UK
Photomask Print	Fineline Imaging Inc., CO, USA
	Advance Reproductions Corporation, MA, USA
	HTA Photomask Inc., CA, USA
	Nanofilm, CA, USA
	Photo Sciences, Inc., CA, USA
	CAD/Art services, OR, USA
Plastic Devices	Ibidi Am Klopferspitz, Germany
PMMA Microchip CE Devices	Microfluidic ChipShop, Germany

Appendix VII : CAS Registry Numbers

Chemical Name	Formula	CAS Number
(3-aminopropyl)triethoxysilane (3-APTES)	$H_2N(CH_2)_3Si(OC_2H_5)_3$	919-30-2
1,10-phenanthroline	$C_{12}H_8N_2$	66-71-7
1,5-diphenylcarbazide (1,5-DPC)	$C_{13}H_{14}N_4O$	140-22-7
4-aminophenol	$H_2NC_6H_4OH$	123-30-8
Acetic acid	CH_3CO_2H	64-19-7
Acetone	CH_3COCH_3	67-64-1
Aluminum (III) sulfate hydrate	$Al_2(SO_4)_3 \cdot xH_2O$	17927-65-0
Ammonium chloride	NH_4Cl	12125-02-9
Ammonium dichromate	$(NH_4)_2Cr_2O_7$	7789-09-5
Ammonium hydroxide	NH_4OH	1336-21-6
Amplex Red	$C_{14}H_{11}NO_4$	119171-73-2
Aniline	$C_6H_5NH_2$	62-53-3
Arginine	$C_6H_{14}N_4O_2$	7200-25-1
Aspartic acid	$C_4H_7NO_4$	617-45-8
Barium (II) chloride	$BaCl_2$	10361-37-2
Bathocuproine	$C_{26}H_{20}N_2$	4733-39-5
Bromocresol green	$C_{21}H_{14}Br_4O_5S$	76-60-8
Cadmium (II) nitrate tetrahydrate	$Cd(NO_3)_2 \cdot 4H_2O$	10022-68-1
Calcium chloride dihydrate	$CaCl_2 \cdot 2H_2O$	10035-04-8
Cerium (IV) ammonium nitrate	$Ce(NH_4)_2(NO_3)_6$	16774-21-3
Chloroform	$CHCl_3$	67-66-3
Chromium (III) chloride hexahydrate	$CrCl_3 \cdot 6H_2O$	10060-12-5
Citric acid	$C_6H_8O_7$	77-92-9
Cobalt (II) chloride	$CoCl_2$	7646-79-9
Cobalt (II) nitrate hexahydrate	$Co(NO_3)_2 \cdot 6H_2O$	10026-22-9

(Continued)

(Continued)

Chemical Name	Formula	CAS Number
Copper (II) sulfate pentahydrate	$CuSO_4 \cdot 5H_2O$	7758-99-8
D (+)-Trehalose dehydrate	$C_{12}H_{22}O_{11} \cdot 2H_2O$	6138-23-4
Dimethylglyoxime (DMG)	$C_4H_8N_2O_2$	95-45-4
Dipotassium hydrogen phosphate	HK_2PO_4	16788-57-1
Ethanol	C_2H_6O	64-17-5
Fluorescein-5-isothiocyanate (FITC)	$C_{21}H_{11}NO_5S$	27072-45-3
Glacial acetic acid	CH_3CO_2H	64-19-7
Glutamic acid	$C_5H_9NO_4$	56-86-0
Glutaraldehyde	$C_5H_8O_2$	111-30-8
Glycine	$C_2H_5NO_2$	56-40-6
Gold (III) chloride trihydrate	$HAuCl_4 \cdot 3H_2O$	16961-25-4
Hydrochloric acid	HCl	7647-01-0
Hydrogen peroxide	H_2O_2	7722-84-1
Hydroxylamine	H_3NO	7803-49-8
Hydroxypropyl cellulose	$C_{36}H_{70}O_{19}$	9004-64-2
Iron (II) chloride tetrahydrate	$FeCl_2 \cdot 4H_2O$	13478-10-9
Iron (III) chloride hexahydrate	$FeCl_3 \cdot 6H_2O$	10025-77-1
Potassium hydrogen phthalate (KHPth)	$C_8H_5KO_4$	877-24-7
Lactic acid	$C_3H_6O_3$	50-21-5
Lead (II) nitrate	$Pb(NO_3)_2$	10099-74-8
Magnesium sulfate heptahydrate	$MgSO_4 \cdot 7H_2O$	10034-99-8
Manganese chloride tetrahydrate	$MnCl_2 \cdot 4H_2O$	13446-34-9
Methanol	CH_3OH	67-56-1
N-(1-naphthyl)-ethylenediamine	$C_{10}H_7NHCH_2CH_2N_2 \cdot 2HCl$	1465-25-4
Sodium carbonate	Na_2CO_3	497-19-8
Nickel (II) nitrate hexahydrate	$Ni(NO_3)_2 \cdot 6H_2O$	13478-00-7
Nickel (II) sulfate hexahydrate	$NiSO_4 \cdot 6H_2O$	10101-97-0
Ninhydrin	$C_9H_6O_4$	485-47-2
Nitric acid	HNO_3	7697-37-2
Nitrous acid	HNO_2	7782-77-6
o-Phenol	C_6H_6O	108-95-2
p-amino benzene sulfonamide	$H_2NC_6H_4SO_2NH_2$	3306-62-5
Paracetamol	$C_8H_9NO_2$	103-90-2
Paraffin wax	C_nH_{2n+2}	8002-74-2
Phenol	C_6H_6O	108-95-2
Phenolphthalein	$C_{20}H_{14}O_4$	77-09-8
Phthalic anhydride	$C_8H_4O_3$	85-44-9
Poly(diallyldimethylammonium chloride) (PDDA)	$(C_8H_{16}ClN)_n$	26062-79-3
Polyethylene glycol (PEG)	$H(OCH_2CH_2)_nOH$	25322-68-3
Potassium dihydrogen phosphate	H_2KPO_4	7778-77-0
Potassium iodide	KI	7681-11-0
Rhodamine B	$C_{28}H_{31}ClN_2O_3$	81-88-9
Salicylic acid	$C_7H_6O_3$	69-72-7
Silver (II) nitrate	$AgNO_3$	7761-88-8

(Continued)

(Continued)

Chemical Name	Formula	CAS Number
Tin (II) chloride dehydrate	$SnCl_2 \cdot 2H_2O$	7772-99-8
Sodium acetate trihydrate	$CH_3COONa \cdot 3H_2O$	6131-90-4
Sodium bicarbonate	$NaHCO_3$	144-55-8
Sodium chloride	$NaCl$	7647-14-5
Sodium citrate dihydrate	$Na_3C_6H_5O_7 \cdot 2H_2O$	6132-04-3
Sodium fluoride	NaF	7681-49-4
Sodium hydroxide	$NaOH$	1310-73-2
Sodium nitrite	$NaNO_2$	7632-00-0
Sodium sulfate	Na_2SO_4	7757-82-6
Sodium tetraborate	$Na_2B_4O_7$	1330-43-4
Succinic acid	$C_4H_6O_4$	1330-43-4
Sulfuric acid	H_2SO_4	7664-93-9
Tetrabromophenol blue	$C_{19}H_6Br_8O_5S$	4430-25-5
Tincture iodine	I_2	7553-56-2
Tris-hydrochloride	$NH_2C(CH_2OH)_3 \cdot HCl$	1185-53-1
Tween-20	NA	9002-89-5
Urea	CH_4N_2O	57-13-6
Vanadium (III) chloride	VCl_3	7718-98-1
Zinc (II) nitrate hexahydrate	$Zn(NO_3)_2 \cdot 6H_2O$	10196-18-6

References

1. Eisenberg A. Keeping a laboratory notebook. *J Chem Educ.* 1982;59(12):1045.

2. Rosenthal LC. Writing across the curriculum: chemistry lab reports. *J Chem Educ.* 1987;64(12):996.

3. Russell DR. *Writing in the Academic Disciplines.* Carbondale, IL: Southern Illinois University Press; 1991.

4. Cai L, Wu Y, Xu C, Chen Z. A simple paper-based microfluidic device for the determination of the total amino acid content in a tea leaf extract. *J Chem Educ.* 2012;90(2):232−234.

5. Wang B, Lin Z, Wang M. Fabrication of a paper-based microfluidic device to readily determine nitrite ion concentration by simple colorimetric assay. *J Chem Educ.* 2015;92(4):733−736.

6. Good laboratory pipetting guide. Thermo Fisher Scientific.

7. Artel I. *Ten Tips to Improve Your Pipetting Technique.* Westbrook, Maine: Artel, Incorporated; 1998.

8. IUPAC. In: McNaught AD, Wilkinson A, eds. *Compendium of Chemical Terminology.* Second ed. Oxford: Blackwell Scientific Publications; 1997.

9. Schneider CA, Rasband WS, Eliceiri KW. NIH Image to ImageJ: 25 years of image analysis. *Nat Methods.* 2012;9(7):671−675.

Index

Note: Page numbers followed by "*f*" and "*t*" refer to figures and tables, respectively.

A

Ablation, 13
Absorbance, 57−58
Access holes, drilling, 12−13
Accuracy, 141, 144, 144*f*
Acid−base titrations on paper, 63
  chemicals and supplies, 64
  design of microfluidic device, 64
  design of paper device for, 65*f*
  experimental procedure, 64−66
    lab report, 66
    solution preparation, 65
    titration, 65−66
    titration paper device, designing and
      fabricating, 64−65
  hazards, 64
Agilent bioanalyzer, 1
Air-displacement pipettes, 140
Amino acid determination, 77, 80*f*
p-Amino benzenesulfonamide, 85
Amperometry, 97−98
Amplex red, 121−122
Anti-bovine serum albumin (BSA), 121−122
Aromatic amines, 109
Average, 145
Azo dyes, 109, 110*f*

B

Bathocuproine (BC), 90−91
Beer's law using a smartphone and
     paper device, 57
  calibration curve, 57
  chemicals and supplies, 59
  experimental procedure, 60−62
    color assay, 61
    data analysis, 61
    image acquisition and analysis, 61
    isolation of starch from potato,
      60−61
    lab report, 61−62
    preparation of solutions, 60

  hazards, 60
  microfluidic device design, 59
Beer−Lambert's law, 57
Blood-plasma separation, 21
Bonding, of glass plates, 13−14, 18
  bonding mechanism, 13−14
  putting fluid reservoirs, 14−15
Bovin serum albumin (BSA), 116
Bromocresol green (BCG) colorimetric assay,
    22−23
Buffer solutions, preparation of, 136−137
Buffering capacity, 136−137
Buffer-oxide etchant (BOE) solution, 12

C

Calibration curve and regression line, 145−146
Caliper LabChip platforms, 2
Capillary electrophoresis, 39
Capillary tubes, 27
CAS registry numbers, 152−154
Chemical wet etching, 12, 17−18
Clostridium botulinum, 83
Commercial flow adapters, 15
Coupling reaction, 109, 110*f*

D

Diazonium-coupling reaction, 109
Diazotization, 83−84, 109, 110*f*
Diffusion, defined, 27, 51
Diffusion coefficient, 51−52
Dimethylglyoxime (DMG), 90
Diphenylcarbazone (DPCO), 91
Droplet microfluidics, 47
  chemicals and supplies, 48
  continuous phase, 47
  design, 48
  dispersed phase, 47
  experimental procedure, 48−50
    droplet generation, 49
    fabrication of device, 49

Droplet microfluidics (*Continued*)
    image analysis and quantitation, 49
    lab report, 49–50
    solution preparation, 48
    hazards, 48
    microdroplets, 47
    microreactors, 47

E

EF separation, 6
Electric double layer, 3
Electrochemical paper microfluidic device,
        design of, 99*f*
Electroosmosis, 3–5, 34
Electroosmotic flow (EOF), 3, 33
    chemicals and supplies, 35
    EOF velocity, 33
    experimental procedure, 35–36
        data collection, 35–36
        lab report, 36
        microfluidic device preparation, 35
        solution preparation, 35
    hazards, 35
    microfluidic device design, 34–35
Electroosmotic mobility, 33–34
Electropherogram, 5*f*, 39
Electrophoretic separation in
        microchannel, 4, 39
    chemicals and supplies, 42
    design of microfluidic device, 40–42
    experimental procedure, 42–45
        data analysis, 44
        electrophoretic separation, 42–44
        lab report, 44–45
        sample preparation, 42
        switching samples in microchip, 44
    hazards, 42
End-point ELISA, 122–123
Enzyme-linked immunosorbent
        assay (ELISA), 121
    chemicals and supplies, 124
    enzyme reaction, 122–123, 122*f*
    experimental procedure, 124–126
        data collection, 125
        ELISA procedure, 125
        image analysis and quantitation, 126
        lab report, 126
        solution preparation, 124
        surface modification of microchannel,
            124–125

hazards, 124
    microfluidic device design, 123–124
Evaluation of chip, 18
Exposure time, 11

F

Flow profile setup step, 40
Flow-based synthesis reaction, 109–110
Flow-focusing, 47
Flow-reactor method, 103–104
Fluid reservoirs, 14
Fluorescein-5-isothiocyanate (FITC), 42, 116
Fluorescence detector, 39
Food colors, 28
4-aminophenol (4-AP), 97–98

G

Glass, properties of, 10*t*
Glass microfluidic device, fabrication of, 9
    access holes, drilling, 12–13
    bonding two glass plates, 13–14
        bonding mechanism, 13–14
        putting fluid reservoirs, 14–15
    chemicals and supplies, 15
    experimental procedure, 16–19
        bonding of two glass plates, 18
        chemical wet etching, 17–18
        evaluation of chip, 18
        lab report, 19
        photolithography, 16–17
    hazards, 16
    microfluidic device design, 15
    photolithographic procedure, 10–12
        chemical wet etching, 12
Glass reservoirs, 14
Glucose assay, 70
Glucose oxidase (GOx), 70
Glutaraldehyde, 116, 122
Gold nanoparticles (AuNPs), synthesis of, 103
    chemicals and supplies, 105
    design of nanoparticle synthesis chip, 105*f*
    experimental procedure, 106–107
        lab report, 106–107
        solution preparation, 106
        synthesis of nanoparticle, 106
    hazards, 105–106
    microchip nanoparticles synthesis setup, 105*f*
    microfluidic device design, 105
    reactions involved in, 104*f*
Griess reagent, 83–84, 137

H

Henderson–Hasselbalch equation, 136–137
Horseradish peroxidase (HRP), 70, 121–122
HPLC-chip/MS system, 2
Hydroxylamine, 89–90

I

ImageJ software, 22, 30, 47, 77, 89, 146–149
Image-processing software, 77, 83, 122
Immunoassay, 122
Inkjet printheads, 1–2
Ionic strength, 34
IR-absorption spectroscopy, 109
Isotropic etching, 12, 12f

J

Jell-O chip, 51–52
    design of, 53f
    fabrication of, 53
    and gelatin mixture, 53

K

Kinetic ELISA (k-ELISA), 122–123

L

Lab notebook, 131
    general guidelines for, 131–133
    sections in, 133
Lab-based companies, 131
Lab-on-a-chip, 1
Laboratory reports, 133–136
    general criteria for grading, 135–136
    sections in, 134–135
Laminar flow, 2
    visualization of, 30
Laminar flow and diffusion in
        microchannel, 51
    chemicals and supplies, 52
    experimental procedure, 53–54
        fabrication of Jell-O Chip, 53
        lab report, 54
        laminar flow setup, 53–54
        solution preparation, 53
    hazards, 52
    microfluidic device design, 52
    mixing, 51
Limit of detection (LOD), 146
Linux, 146
Local surface plasmon resonance (LSPR), 103

M

Mac OS X, 146
Mean, defined, 145
Metal ions, colorimetric determination of, 89
    chemicals and supplies, 92
    colorimetric assays, 89–91
    experimental procedure, 93–94
        determination of metal ions, 93–94
        fabrication of device, 93
        lab report, 94
        preparation of solutions, 93
    hazards, 93
    microfluidic device design, 92
    μPAD approach, 89
    quantitative analysis, 89
Metal pollution, 89
Methemoglobinemia, 83
Microchip electrophoresis systems, 39
    design of, 40–42
    hazards, 42
Microfabrication, 15
Microfluidic paper analytical device
        (μPAD), 21, 63, 70, 77, 83
    for blood-plasma separation, 22–23, 23f
Microfluidics, defined, 1–2
Micropipettes, 138–143
    accuracy, 141
    calibration of, 141–143
    error in, 141
    guidelines to proper use of, 138–140
    parts of, 139f
    precision, 141
    tips for use, 140–141
    working principle of, 140
Microreactor flow synthesis, 103–104
Micrototal analysis systems, 1
Molar absorption coefficient, 57
Molecular diffusion coefficient, 51
μ-TAS, 1

N

Nanoparticles (NPs), synthesis of, 103–104, 106
Ninhydrin solution, 77, 79
    preparation of, 136
Ninhydrin test
    design of paper device for, 78f
    reaction involved in, 78f
Nitrite ions in water, determination of
    chemicals and supplies, 85
    experimental procedure, 85–87

Nitrite ions in water, determination of
        (*Continued*)
    determination of nitrite ion, 87
    fabrication of device, 86
    lab report, 87
    sample preparation, 85–86
    hazards, 85
    microfluidic device design, 85
    using paper analytical device, 83
Nitrosamines, 83

O
Occupational respiratory diseases, 89
Organic dye, flow synthesis of, 109
    chemicals and supplies, 111
    design of microfluidic device, 110–111
    experimental procedure, 111–112
        dye synthesis, 112
        lab report, 112
        solution preparation, 111
    hazards, 111
Organic solvents, 14

P
P81, 97–98
Paper microfluidics, for blood-plasma
        separation, 21–22
    chemicals and supplies, 24
    device design, 23
    experimental procedure, 24–25
        fabrication of the device, 24
        lab report, 25
        separation procedure, 24–25
        solution preparation, 24
    hazards, 24
    protein assay, 22–23
Paper-based Griess method, 84
Paracetamol (PA) and 4-aminophenol mixture,
        analysis of, 97
    amperometry, 97–98
    chemicals and supplies, 98
    electrochemical detection, 97–98
    experimental procedure, 99–100
        electrochemical measurements, 100
        fabrication of the device, 99
        lab report, 100
        preparation of solutions, 99
    hazards, 99
    microfluidic device design, 98
    three-electrode system, 97–98

Péclet number, 51–52
Permanent-marker method, 59
Personal protective equipment, 111
Phenols, 109
Phosphate buffer, preparation of, 137
Photolithography, 10–12, 16–17, 16*f*
Photomask, 11, 15*f*
Photomultiplier tube (PMT) detection, 39
Photoresist, 10–11
Pinched injection scheme, 39–40, 40*f*
Pipette tips, 14
Piranha solution, 14
Plasma ashing, 13
Point-of-care blood analyzers, 2
Poly(acrylic acid), 89–90
Polydimethylsiloxane (PDMS) microchip, 27
    chemicals and supplies, 28
    design of, 27
    experimental procedure, 28–31
        fabrication, 28–30
        lab report, 30–31
        visualization of laminar flow, 30
    hazards, 28
Positive-displacement pipettes, 140
Potassium hydrogen phthalate, 63
Potato, starch isolation from, 60–61
Precision, 141, 144, 144*f*
Pressure-driven flow, 52
Protein and glucose in urine sample,
        simultaneous determination of, 69
    chemicals and supplies, 71
    experimental procedure, 72–74
        bioassays, 73
        fabrication of device, 72
        lab report, 73–74
        solution preparations, 72–73
    glucose assay, 70
    hazards, 71
    microfluidic device design, 71
    protein assay, 69
    quantitative analysis, 70
    undergraduate lab, 70
Protein assay, 22–23, 69
Protein immobilization on glass microfluidic
        channel, 115
    chemicals and supplies, 117
    experimental procedure, 117–119
        data collection, 118
        image analysis and quantitation, 118–119
        lab report, 119

solution preparation, 117
  surface modification of microchannel, 118
hazards, 117
microfluidic bioassay methods, 115
microfluidic device design, 117
physical adsorption methods, 115
Proteinuria, 69

R
Recordkeeping, 131–136
  lab notebook, 131
    general guidelines for, 131–133
  laboratory reports, 133–136
    general criteria for grading, 135–136
Red blood cells (RBCs), 21
Regression line, 145–146
Reservoirs, 14–15, 63, 92
Reynolds number, 2, 51
RGB (red, green, blue) color model, 58
Rhodamine B, 15, 35
Room temperature (RT) direct bonding, 13
Room-temperature bonding procedure, 18*f*

S
Sandwich ELISA, 121
Scientific papers, 134
Separation resolution, 5–6
Service provider information, 149–152
Sodium hydroxide solution, 63
Sodium nitrite, 83
Standard addition method, 84
Standard deviation (SD), 141, 145
Statistical treatment of data, 144–146
  average and standard deviation, 145
  calibration curve and regression line, 145–146
  limit of detection (LOD), 146
Straight-channel design, 15, 15*f*
Sulfuric acid, 14
Surface plasmon resonance, 103–105

T
Tetrabromophenol blue (TBPB), 69, 71
(3-aminopropyl)triethoxysilane (3-APTES),
  116, 118, 122
Titrations, 63
T-junction microfluidic device, 47
  fabrication of, 49
Total amino acids in tea, quantitative
    determination of, 77
  chemicals and supplies, 78–79
  experimental procedure, 79–81
    determination of amino acids, 79–80
    fabrication of device, 79
    lab report, 80–81
    preparation of reagents and solutions, 79
    tea-leaf extraction, 79
  hazards, 79
  microfluidic device design, 78

U
Urinalysis, 69
UV–vis absorption spectroscopy, 109
UV–vis spectroscopy, 104

V
Vendors and service provider information,
  149–152
Viscous forces, 51

W
Water, 14
Wax dipping, 21
Wax-pen method, 59
Wax-printed μPADs, 63
Wax-printing method, 59
  for paper-device fabrication, 63–65, 72
Whatman No. 1 chromatographic paper, 97–98
Whitesides Group of Harvard University, 21
Windows, 146

Printed in the United States
By Bookmasters